# Self-Reconfigurable Robots

**Intelligent Robotics and Autonomous Agents**
Edited by Ronald C. Arkin

For a complete list of the books published in this series, please see the back of this book.

# Self-Reconfigurable Robots

## An Introduction

**Kasper Stoy**
**David Brandt**
**David J. Christensen**

The MIT Press
Cambridge, Massachusetts
London, England

**4    Electrical Design of Self-Reconfigurable Robots**                    **83**
    4.1    Computing and Communication Infrastructure              83
    4.2    Energy                                                  89
    4.3    Sensors                                                 90
    4.4    Conclusion                                              91
    4.5    Further Reading                                         93

**5    The Self-Reconfiguration Problem**                                   **95**
    5.1    Formulating the Problem                                 95
    5.2    Why Is the Self-Reconfiguration Problem Difficult?      97
    5.3    Simplifications of the Self-Reconfiguration Problem     102
    5.4    Conclusion                                              110
    5.5    Further Reading                                         111

**6    Self-Reconfiguration as Search**                                     **113**
    6.1    Configuration Representation                            114
    6.2    Search Space Considerations                             115
    6.3    Informed Search                                         117
    6.4    A Successful Search Requires Simplifications            121
    6.5    From Solution to Control                                122
    6.6    On-Line Distributed Search                              123
    6.7    From Impossible to Simple                               123
    6.8    Further Reading                                         124

**7    Self-Reconfiguration as Control**                                    **127**
    7.1    Movement Strategy                                       128
    7.2    Representation of the Goal Configuration                134
    7.3    Complications                                           138
    7.4    Docking and Merging                                     139
    7.5    Making Ends Meet                                        140
    7.6    Further Reading                                         140

**8    Task-Driven Self-Reconfiguration**                                   **145**
    8.1    Locomotion through Self-Reconfiguration                 146
    8.2    Task-Driven Growth                                      147
    8.3    Self-Reconfiguration as a Side Effect                   151
    8.4    New Challenges in Self-Reconfiguration                  152
    8.5    Conclusion                                              154
    8.6    Further Reading                                         155

**9    Control in Fixed Configurations**                                    **157**
    9.1    Locomotion                                              157
    9.2    Manipulation                                            169

9.3 Conclusion 170
9.4 Further Reading 170

**10 Research Challenges** **173**
10.1 Facing the Complexity of Real Tasks 174
10.2 From Basic Functionalities to Behaviors 175
10.3 Behavior Adaptation 176
10.4 Behavior Selection 177
10.5 Behavior Mode 178
10.6 Behavior-Based Robotics as a Framework 179
10.7 Application-Oriented Hardware 179
10.8 Conclusion 180
10.9 Further Reading 180

**Appendix: A Simulator for Self-Reconfigurable Robots** **183**

References 185
Index 193

# Foreword

Wouldn't it be great if we could not only change the shape that our arms or legs bend into, but disconnect and reconnect them in different places? Add extra arms or lengthen them? Self-reconfigurable robots aim to do something like that, changing their physical connectivity to suit a task. This idea is catching on. If you watch movies or read science fiction, there are countless examples of morphing robots, liquid-metal robots, nanorobots, and the like. While in science fiction self-reconfigurable robots are often world-conquering, with evil intent, this book presents research on self-reconfigurable robots that is growing worldwide, attracting active researchers and gaining public interest.

With this growth it is somewhat surprising that a book that explores this phenomenon has not yet been written. So the writing of this book is certainly timely; that Kasper Stoy, David Brandt, and David Christensen have done it in such wonderful fashion is especially gratifying to a person who has been active in the field for many years. This book collects and distills two decades' worth of research in a cogent fashion. It will be useful to students hoping to develop projects or thesis topics and even those robot movie enthusiasts who wonder how far we are from seeing a liquid-metal robot. However, most directly, this book will be valued by the self-reconfigurable robot research community.

This research community started with a handful of scientists and academics in the early 1990s and has grown to many dozens of groups located all over the world. The community has several opportunities each year to gather to discuss its members' research. In these conferences and workshops, researchers present their latest accomplishments in a positive light and talk about future directions and promises of their development. People are almost always congenial and congratulatory. It was in one of the many workshops held by this community that Kasper made his mark on me and set the tone for his research, and I think this book as well.

In the early 1990s I said that modular self-reconfigurable systems held three promises:

• Some of the remaining research challenges of self-reconfigurable robots that we need to address in order to realize the full potential of these robots

This book is targeted at students at the graduate level and researchers interested in the field of self-reconfigurable robots. It can work as a course book as well as for self-study. If it is used as a course book, for the topics a teacher finds most important, he or she can supplement the book with the papers referenced in the further reading sections. The book is written to appeal to and be accessible to a wide audience. It requires only some basic knowledge of search, and as such, it may also be useful for people who are not part of the core audience but who want to become aquainted with the field of self-reconfigurable robots.

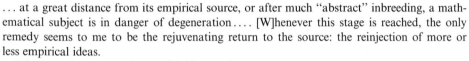

... at a great distance from its empirical source, or after much "abstract" inbreeding, a mathematical subject is in danger of degeneration.... [W]henever this stage is reached, the only remedy seems to me to be the rejuvenating return to the source: the reinjection of more or less empirical ideas.

—John von Neumann, *Theory of Self-Reproducing Automata*

**Figure 1.4**
Many different robots can be built by connecting modules in different ways. Here a number of ATRON modules are connected to form snakelike robots as shown on the left and a small, carlike robot shown on the right.

**Figure 1.5**
A self-reconfiguration step in the ATRON robot consists of a disconnect, a rotation, and a connect operation.

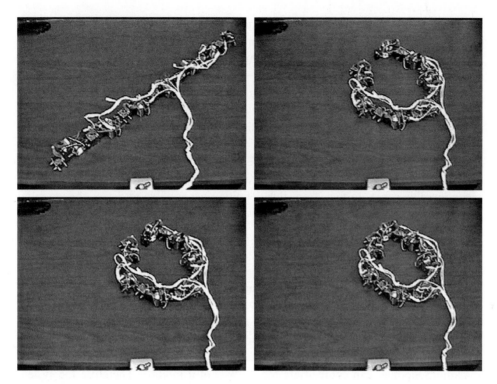

**Figure 1.6**
A self-reconfiguration step in the CONRO robot. (a) The robot starts in a long, snakelike configuration. (b) It then bends in on itself. (c) The two ends search and locate each other using infrared sensors and transmitters. (d) The two ends connect: the right module connects to a side connector of the left module. (Courtesy of Shen)

*Reconfigurable*   The modules can be connected in several different ways to form different robots in terms of size, shape, or function.

*Dynamically reconfigurable*   The modules can be disconnected and connected while the robot is active.

*Self-reconfigurable*   The robot can change the way modules are connected by itself.

The nature of being modular is to encapsulate some of the complexities of the functionality of a module. This means that while regular screws are not modules, a drilling machine is. However, in order to be part of a modular robot, a drilling machine has to be connected to at least another module, e.g., a robot arm.

Our example drilling robot is not a reconfigurable robot. However, if we extend it with more tool modules, which can replace the drilling machine, then it becomes reconfigurable. If, furthermore, we are able to change modules while the robot is active, it becomes a dynamically reconfigurable robot. For example, if we can replace the tools of our example robot, it becomes dynamically reconfigurable. Finally, if the robot can change by itself the way modules are connected, it becomes a self-reconfigurable robot. That is, if our example robot arm can change its tools, it is a self-reconfigurable robot.

The example robot arm with replaceable tools represents an extreme of what defines a self-reconfigurable robot. However, it is clear that it is not a typical self-reconfigurable robot. A typical self-reconfigurable robot is extendible with regard to the number of modules; the complexity of different modules is often comparable; and the number of modules is often higher. We could have created a more restrictive definition, but the proposed definition serves as a good guideline for what defines a self-reconfigurable robot. In fact, from this definition we are able to derive the potential features of self-reconfigurable robots, as we will see in the following section.

## 1.2 Features

Self-reconfigurable robots have some unique features that make them interesting from an engineering point of view. Owing to their modular nature, self-reconfigurable robots have a high degree of redundancy, which they can exploit to become *robust*. A hardware failure or a software error may cause a module to fail. This, however, does not cause the self-reconfigurable robot as a whole to fail. The remaining modules can compensate for the loss of a module. Therefore the system is robust and its performance degrades gracefully with the number of failed modules. It may also be possible for the robot to use its ability to change shape to replace broken modules with spare ones in the system if any exist. Through this self-repair ability, the robot may stay functional even if a substantial number of modules fail.

A self-reconfigurable robot is *versatile*: the modules can be combined in many different ways, allowing them to form the basis for a wide range of different robots. Furthermore, a self-reconfigurable robot is *adaptable* because it can continually adapt and even completely change shape if a task requires it.

Self-reconfigurable robots are *cheap compared with their complexity*. The individual modules are quite complex and, as such, are expensive to produce. However, a self-reconfigurable robot consists of many identical modules and therefore, the cost of the individual module can be lowered because they can be mass produced.

In summary, self-reconfigurable robots are versatile, adaptable, robust, and cheap considering their complexity. It is important to note that these features are only potential features. In theory, it should be possible to realize these features based on the concept of self-reconfigurable robots. However, in practice, the features are often realized only to a limited degree, as we will see in later chapters. Before we go into the technical details, let us put these robots into context by taking a tour through their history.

## 1.3 Brief History

The hope of creating robust, versatile, adaptable, and cheap robots has led researchers to develop a succession of ever-improving self-reconfigurable robots. In this section we review the history of self-reconfigurable robots so we can understand and appreciate the advances that have resulted in the modern generation of these robots.

### 1.3.1 From the Industrial Revolution to Robot Manipulators

Self-reconfigurable robots and robots in general are probably best understood in the context of the Industrial Revolution. The Industrial Revolution started around 1733 in the textile industry with inventions such as Sir Richard Arkwright's spinning frame, the first automated spinning machine. During the eighteenth century, many other machines were invented and the revolution picked up pace. It was further reinforced by James Watt's steam engine, introduced in 1769, which replaced water as a power supply.

After the invention of the worm gear, introduced by Ramsdan's dividing engine in 1774, the precision of machines was increased and spread to other industries. Improved precision also meant that the concept of replaceable parts became possible because the machines now had high enough precision to make things to specification every time. This opened the path to mass production, which was refined during the nineteenth century to culminate in Henry Ford's car factory in 1908, which pioneered the first conveyor belt assembly line.

On the assembly line, tasks were cut into small simple tasks so that an unskilled worker could do quickly and efficiently. The simplification of these tasks made it possible to introduce the next generation of machines—the robot manipulators. General Motors introduced the Unimation 1900 into their car assembly line in 1961. Unimation Inc., founded by George Devol five years earlier, produced this robot. In the following years, and even today, robot manipulators are optimized for doing one task fast and precisely. The result of this is the incredibly fast and precise robot manipulators we have today. The ABB IRB340 FlexPicker, shown in figure 1.7, has a top speed of 10 m/s, a precision of 0.1 mm, and can carry a payload of 2 kg. Other trade-offs between speed and payload exist, such as the KUKA KR 6/2, which can handle a payload of 6 kg and a top speed of 4 m/s.

### 1.3.2 Robots in Fiction

In 1920 the author Josef Capek published his play "Rossum's Universal Robots," in which he described machine slaves, i.e., robots, that would aid humanity. Little did he know that this idea would form the guiding light and inspiration for many engineers and scientists for the rest of the twentieth century and into the twenty-first. The technology at the time was of course not mature enough to even attempt to realize

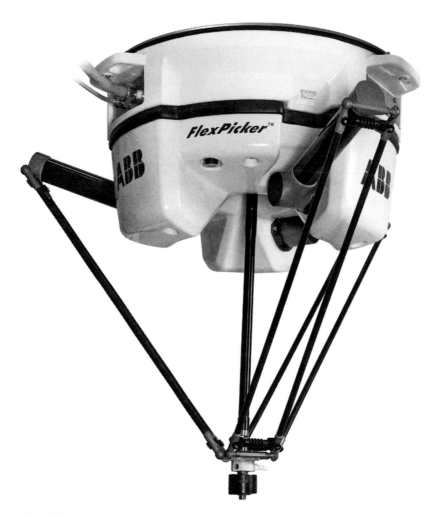

**Figure 1.7**
The IRB 340 FlexPicker. This is the culmination so far in the continuous effort of creating faster and more precise robot manipulators. (Courtesy of ABB, © 2005)

this dream; rather, mechanical multiplication machines were at the cutting edge in the 1930s.

Isaac Asimov was the next to elaborate on robots with his famous short stories "Robbie" in 1940 and "Runaround" in 1942 (both short stories can be found in the 1950 book titled *I, Robot*). Asimov's writing on robots mainly considered the ethical aspects of robots and he also formulated the now-famous three laws of robotics (from "Runaround"):

1. A robot may not injure a human being or, through inaction, allow a human being to come to harm.

2. A robot must obey orders given to it by human beings except where such orders would conflict with the First Law.

3. A robot must protect its own existence as long as such protection does not conflict with the First or Second Law.

After this, robots were a recurring theme in science fiction. However, for self-reconfigurable robots, the big breakthroughs were in the movies *Terminator 2: Judgment Day* (1991), *The Matrix Revolutions* (2003), and the series of transformers comic books and movies, culminating in the movie *Transformers* (2007). All these movies featured robots that could automatically change their own shapes. Perhaps some of these works of fiction inspired Toshio Fukuda, who among others, created the philosophical foundation for the field of self-reconfigurable robots as we know it today.

### 1.3.3  The Cellular Robot

In the 1980s the idea of distributed robotic systems emerged. The idea was that instead of building robots as monolithic, inflexible pieces of hardware optimized for speed and precision, they could be built using a cellular design not unlike the one employed by nature. These cellular robots could autonomously split into their constituent cellular, robotic modules that later could recombine to form a new robot. One example Toshio Fukuda gave was a robot that could move into environments that are difficult to reach, e.g., accessible only through narrow openings, and once inside change shape by itself to accomplish a task (see figure 1.8). The hope was that this approach would provide robots with an unprecedented level of versatility and robustness.

The early work was mainly conceptual in nature because the technology was not mature enough at the time to realize many of the ideas. For example, in 1991 G. Beni and J. Wang [4, 47] concluded, after having worked in the area for a while, that:

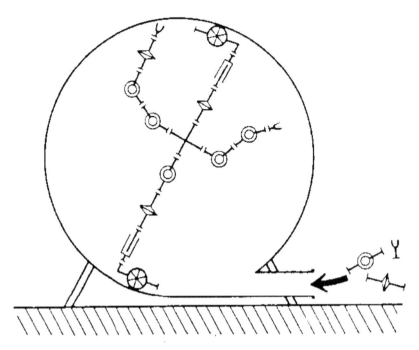

**Figure 1.8**
An early artistic impression by Fukuda [45] of cellular robots doing maintenance work in a storage tank.
(Courtesy of Fukuda, © 1988 IEEE)

While the most urgent problem for the realization of distributed robotic systems is the construction of the physical structure of the units, the most fundamental problem remains the design of algorithms. [4: 1919]

One of the people who started to implement part of the conceptual framework was Toshio Fukuda. He began by considering which types of modules were needed to build a robot. In his original work he proposed a heterogeneous system consisting of three types of modules. Type 1 modules would be used for actuation, i.e., joints and wheeled modules. Type 2 modules would be structure modules, i.e., branching modules or power modules. Type 3 modules would be modules with tools and other special-purpose modules. He was also interested in how they could be combined to accomplish a task. In the following years Fukuda worked toward realizing this vision through many iterations of the CEBOT system (CEllular roBOT), which in essence became a multirobot system consisting of mobile robots [44].

In this early work, self-reconfigurable robots were not as clearly defined as today since the distinctions between other types of robots such as multirobot systems, sensor networks, or cyborg, were not made yet. Instead they were all considered distrib-

**Figure 1.9**
A segment module of the PolyPod robot, a predecessor of chain-type self-reconfigurable robots. (Courtesy of Yim, © 1994)

uted robotic systems. However, in this early beginning, the motivation for creating self-reconfigurable robots, which still drives us today, became clear.

### 1.3.4 The First Self-Reconfigurable Robots

The first self-reconfigurable robots became reality a few years later. In 1993 Mark Yim [139] built PolyPod, which demonstrated the versatility of a modular design (see figure 1.9). Yim showed that by connecting the two types of modules of the Poly-Pod in different ways, he could implement many different gaits [140]. The PolyPod robot was dynamically reconfigurable, but was unable to change shape by itself and therefore is not a self-reconfigurable robot. Mechanically, PolyPod is an important predecessor to the class of chain-type self-reconfigurable robots we introduced in section 1.1 and that we will describe in more detail in section 3.1 in chapter 3. However, for now, the important aspects of chain-type robots are that they form chains and treelike structures, possibly also containing loops, and are mainly geared toward locomotion.

Around the same time, another two new robots, shown in figure 1.10, arrived: the Fracta robot built by Satoshi Murata, Haruhisa Kurokawa, and Shigeru Kokaji [73] at the National Institute of Advanced Science and Technology in Tsukuba, Japan and the Metamorphic robot built by Gregory Chirikjian [27] at Johns Hopkins University, in Baltimore, Maryland. These robots demonstrated the ability to change shape in two dimensions. Interestingly enough, the self-reconfiguration ability was implemented in two completely different ways. In Fracta it was realized using electromagnets, and in the Metamorphic robot it was realized mechanically. In these

**Figure 1.10**
Early lattice-type self-reconfigurable robots: Top: Three Fracta modules. (Courtesy of Murata) Bottom: Two modules of the Metamorphic robot. (Courtesy of Chirikjian)

robots the configuration of modules forms a lattice, which is why they later became known as lattice-type self-reconfigurable robots, which, as mentioned, are robots in which the modules are organized like atoms in a crystal.

### 1.3.5 The Exploration Phase

In these early years the split between chain-type and lattice-type self-reconfigurable robots was made. In the following years, the chain-type systems were improved to be able to do self-reconfiguration, and the lattice-type systems were improved to achieve three-dimensional self-reconfiguration.

A new chain-type self-reconfigurable robot, CONRO [24], was introduced in 1999 by Andres Castaño et al. [24] and the following year Yim et al. [142] presented an improved version of PolyPod, the PolyBot robot (both shown in figure 1.11). Although both systems excelled in demonstrating a wide range of locomotion patterns, both systems struggled to demonstrate self-reconfiguration. It turned out to be more difficult than expected to make two chains meet and connect based on infrared sensing alone. A systematic solution to this problem has actually been proposed recently [89], but it remains a significant problem.

Self-reconfiguration in three dimensions was also a difficult problem, owing to gravity and geometrical constraints. In three dimensions, modules have to be strong enough to lift other modules against gravity to achieve self-reconfiguration. In two dimensions, the third dimension is unconstrained and it was used as extra space and was an obvious place to attach wires for a power supply. This was no longer possible in three dimensions; the complete module had to fit into the lattice structure. The earlier Fracta and Metamorphic robots had used a method of "rolling around" neighbors using tracks running around modules. It was not obvious how that generalized to three dimensions because the "tracks" then would have to cross each other. It was therefore a breakthrough when in 1997, Keith Kotay and Daniella Rus [95], then at Dartmouth College in Hanover, New Hampshire, presented the Molecule and Murata et al. [75] presented the 3D-Unit, the first three-dimensional self-reconfigurable robots (see figures 1.12 and 1.13). The solution adopted was an internal actuator in a module that allowed one part of a module to rotate with respect to the other. In the earlier systems, moving and connecting were the same physical action. However, in the Molecule for the first time these actions were split into a sequence consisting of a disconnection, a move, and a connection. This split the self-reconfiguration problem into, on one side, connecting and disconnecting and, on the other, moving, and thus making the mechanical problem more manageable. Ünsal et al. [129] at Carnegie Mellon University in Pittsburgh, Pennsylvania, also used a similar idea in the I-Cubes system. This system was heterogeneous and consisted of passive cubes and active links.

Another implementation of self-reconfiguration is to make modules contract and expand rather than rotate. This approach was pioneered in two dimensions in Rus and Vona's [97] Crystalline robot in 1999 and in three dimensions in Suh et al.'s [122] Telecube in 2001, shown in figure 1.14. A self-reconfiguration step would then typically consist of an expansion, a connection to a module in a neighboring lattice position, a disconnection from neighboring modules in the original lattice position, and a contraction. The contraction and expansion were implemented using telescoping arms, which is a mechanically fragile solution.

An attempt was also being made to miniaturize the two-dimensional Fracta, the result of which was the Micro-Unit developed by Eiichi Yoshida et al. [150] at the

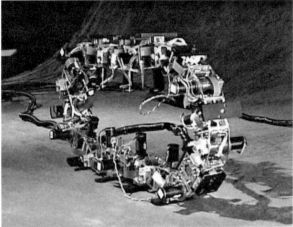

**Figure 1.11**
Chain-type self-reconfigurable robots. Top: CONRO in a hexapod configuration. (Courtesy of Stoy, © 2002 IEEE) Bottom: Second-generation PolyBot in a loop configuration. (Courtesy of Yim, © 2000 IEEE)

**Figure 1.12**
Top: A module of the Molecule self-reconfigurable robot. (Courtesy of Kotay, © 2005 IEEE) Bottom: A configuration of four modules. (Courtesy of Rus) The Molecule robot was the first self-reconfigurable robot able to perform self-reconfiguration in three dimensions.

National Institute of Advanced Industrial Science Art Technology, Tsukuba, Japan. The final prototype of the Micro-Unit [151] was very small, with a volume of 8 cm$^3$, excluding electronics.

Toward the end of this period the two branches of self-reconfigurable robots had matured. Chain-type and lattice-type robots complemented each other: chain-type robots were able to produce advanced locomotive gaits and lattice-type robots could change shape in three dimensions.

### 1.3.6 The Hybrids

In 1999 the two branches of lattice-type and chain-type self-reconfigurable robots were merged in the M-TRAN robot by Murata et al. [76] (see figure 1.15). Owing

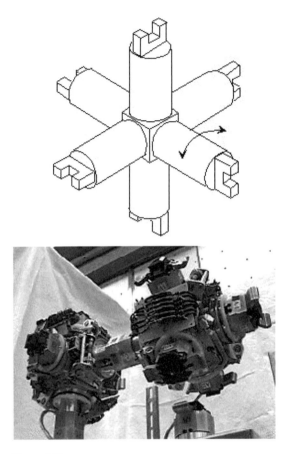

**Figure 1.13**
The 3D-Unit module. Top: The basic design of a single module; each module has six connectors and six rotational joints. Bottom: A picture of two connected 3D-Unit modules. (Courtesy of Murata)

to an innovative mechanical design, an M-TRAN could exist in both a lattice structure, making self-reconfiguration relatively easy, and in a chain-type structure, making locomotion easy. In 2006 Shen et al.'s [103] SuperBot appeared, which included an extra degree of freedom compared with the M-TRAN robot. In M-TRAN the actuators were parallel to each other; in SuperBot a degree of freedom was added to make the orientation between these two actuators controllable. The ATRON robot developed in 2003 by Jorgensen et al. [52, 82] at the University of Southern Denmark, Odense, was the second hybrid robot. It introduced the novel idea that three-dimensional self-reconfiguration could be achieved even though each module only had one actuator. This was accomplished by arranging modules, and thus their rotational axes, perpendicular to each other. Victor Zykov et al. [159] at Cornell University in Ithaca, New York, used a similar idea in the design of the Molecube robot.

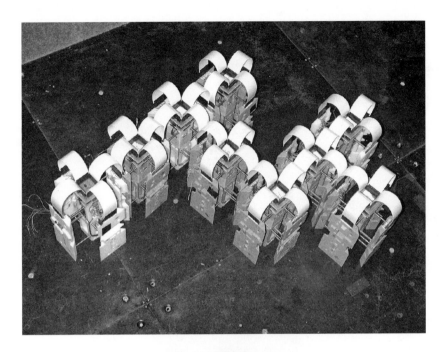

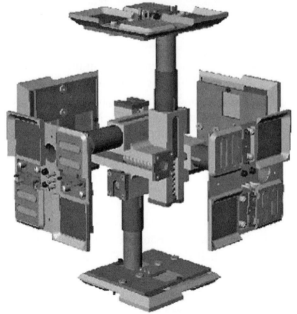

**Figure 1.14**
The Crystalline robot (top, courtesy of Rus) and the Telecube (bottom, courtesy of Suh, © 2002 IEEE).
The modules of these robots are able to contract and expand.

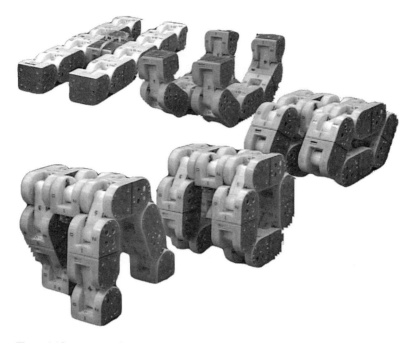

**Figure 1.15**
The M-TRAN robot combined the features of lattice-type and chain-type robots in one system through an
innovative mechanical design. (Courtesy of Murata)

In the meantime, other systems were also created to improve on earlier designs.
There is the Gear-Type unit introduced by Hiroko Tokashiki et al. [126] at the University of Ryukyus, Japan, in which modules are magnetic gears and thus can roll
around each other quickly, and Chobie, developed by Michihiko Koseki et al. [59]
at the Technical University of Tokyo, Japan, which is a vertical, two-dimensional
module. However, like the early designs, it is not obvious how to generalize these to
three dimensions.

### 1.3.7   State of the Art

Self-reconfigurable robots have undergone almost twenty years of development and
today we have several self-reconfigurable robots that can reliably perform different
locomotion patterns and change their own shape. This essentially means that the
basic technical challenges that prevented the early thinkers from realizing their vision
have been overcome. However, new challenges await the research community as self-
reconfigurable robots leave the realm of basic research and move toward application.
We postpone the discussion of these challenges to chapter 10, at which point we will
have gained a deep insight into the design and control of these robots and thus a better basis for understanding the challenges.

## 1.4  Pack, Herd, and Swarm Robots

Self-reconfigurable robots vary in terms of how many modules it takes to construct a robot and how small the individual modules are. Here we define three categories of self-reconfigurable robots. Even though they are not standard categories in the literature, we have found them to be useful in this book. The categories are pack, herd, and swarm robots.

*Pack robots* consist of tens of modules. The modules are generally characterized by having a strength comparable to the group's size. This means that the individual module by itself is useful and certainly can lift itself, but also is able to lift a large fraction of the other modules in the robot and thus make it possible for one module to be a functional unit in the robot, such as a leg. Each module in the system plays a crucial role and it is therefore of crucial importance that they are strictly coordinated to work together to achieve the robot's goal. We can compare this to a pack of wolves hunting, where the performance of each wolf is significant to the outcome of the hunt. We therefore refer to these robots as pack robots.

*Herd robots* consist of hundreds of modules. The strength of these modules is moderate compared with the group size, and the functionality of the individual is limited. This means that one module is still able to lift itself but cannot be a functional unit in the robot by itself. Functional units are always built from groups of modules. In these systems there is enough redundancy to allow less strict coordination without significantly affecting the performance of the system. It is still possible to control each module centrally, but at a significant cost in performance. In these robots the modules generally work together to perform the robot's task, but not as tightly as in pack robots; modules can stray from time to time. Modules in these robots can be compared to deer in a herd and thus we call these robots herd robots.

*Swarm robots* are a more common term. These robots consist of myriads of modules. Individual modules are weak and have limited influence on the robot as a whole. Only by coming together can the modules do something that influences the robot, and massive numbers of modules are needed to create a functional unit in the robot. In these robots it is impossible to control the myriads of modules centrally and therefore the modules must be autonomous to a high degree. These robots act as swarms that cannot be controlled, but live and develop according to their own rules, similar to swarms of bees or ants. We therefore refer to these robots as swarm robots. To summarize:

*Pack robots*   These robots consist of tens of modules and must usually be tightly coordinated because the actions of individual modules are crucial for the performance of the robot.

*Herd robots*   These robots consist of hundreds of modules and can be globally coordinated only with difficulty; they are better controlled as a collection of groups since the actions of individual modules are still important but not crucial for the performance of the robot.

*Swarm robots*   These robots consist of myriads of modules and, owing to the number of modules, are difficult to coordinate globally. Instead each module is controlled locally, which is possible because each module has little effect on the overall behavior of the robot.

The rationale for having a herd category requires a little more explanation. The main point is that robots in the herd category are problematic. Algorithms designed for pack robots that are either centralized or require modules to be tightly coupled typically start to face scaling problems: it is difficult to keep hundreds of modules tightly coordinated, particularly for robots relying on local communication. On the other hand, it is difficult to scale down the highly scalable algorithms for swarm robots because the movement of each module is important for the herd as a whole and therefore the stochastic processes of swarm robots are not well suited. In other words, you still need the tight coordination of pack robots even though the robot is medium sized. It may be possible to break this barrier between pack robots and swarm robots by organizing the system in a hierarchy, but for now we view herd robots as problematic.

The classification described here does not apply only to hardware. We can also talk about pack controllers, herd controllers, or swarm controllers. This distinction is important because on the one hand, in the pack controllers randomness cannot be used since each module has to be controlled carefully; on the other hand, in the swarm controllers randomness is a powerful mechanism and it is impossible to control each module carefully (i.e., provide it with all the information it needs to select the optimal action). We find this distinction useful for classification of systems and it can help a researcher decide when to apply which algorithm. For example, we cannot hope to apply pack algorithms to swarms and the other way around, although there may exist smaller groups of modules within the swarms that we want to control as packs or herds. This is one of the challenges of controlling self-reconfigurable robots: how can we maintain tightly coordinated control in critical parts of the robot while allowing larger parts of the robot to work as a swarm with looser coordination?

## 1.5   From Vision to Application

Self-reconfigurable robots have not been used yet since it is only within the past few years that they have matured to a degree where applications are possible. However, in the twenty years since the idea of self-reconfigurable robots was conceived, numerous applications have been envisioned; they range from being realizable today to pure science fiction. At the immediately realizable end are advanced robot applications that benefit from the unique features of self-reconfigurable robots. At the futuristic end, the idea that self-reconfigurable robots are universal robots, in the sense that they can simulate any robot, is taken to its extreme.

Pack robots are the ones closest to application. They have the advantage that a relatively limited and therefore affordable number of modules are used. The coordination of a limited number of modules is also significantly easier, and solutions to basic tasks such as locomotion and self-reconfiguration exist. Therefore, applications for these types of robots are technologically within reach; the question is whether there is a niche in the marketplace in which pack robots will fit.

Pack robots are generally well suited for exploration and inspection applications. In these applications the pack robot can exploit its versatility to adopt a locomotion style that fits the different environments and terrains it encounters. Locomotion styles may include swimming, running, climbing, and rolling. One proposed application for pack robots is to assist in search and rescue in collapsed buildings. The pack robot could search the building using its ability to change shape to gain access to places human rescue workers cannot reach. Another, similar application from a technical point of view, is sewer inspection. The pack robot becomes even more useful in applications where it can take advantage of its robustness and ability to self-repair. One such application is exploration of extraterrestrial environments. In such hard-to-reach environments it is important that the robot is able to maintain some level of functionality even if some modules fail. A concrete example is the SuperBot project in which the ambition was to support life on other planets [99]. The SuperBot robot would land on another planet, find a suitable location to plant a seed, and finally protect the seed from the environment in the early phases of its life.

The pack robots are small and agile and this gives them advantages in exploration and inspection applications, but as soon as the robot needs to interact with the environment, it needs to be larger and stronger. Therefore, herd robots are often better suited than pack robots for tasks that require interaction with heavy objects. Herd robots, for instance, would be better suited for search and retrieval tasks than pack robots: a herd robot could search a cave for interesting rock samples and transport them back to a lab for further analysis, or it could find and retrieve humans from a collapsed building. The herd robot could also reinforce a collapsed building to make it safe for rescue workers to enter. Herd robots have a sufficiently high number of modules to make it possible to differentiate functionalities of different parts of the robot. We have therefore proposed to make a *morphing production line*.

This production line could handle many of the handling operations in industry today, such as transporting, sorting, manipulating, and assembling objects. A herd robot could take advantage of its adaptability by adapting its shape to the objects being handled, making categorization and sorting easy, and it could also change configuration to match changing demands on the production line. For instance, a herd robot could decide to perform time-intensive tasks in parallel. Another task is creating ergonomic furniture. Here again the idea is to use the adaptability of the herd robot to fit a piece of furniture to the person using it and the task that person is doing. It may even be possible to use furniture as a highly adaptable user interface.

Realizing swarm robots is still an ongoing basic research effort, but if it is successful, the potentials are enormous, depending on the cost. We may inject swarm robots into a blood vessel and have them perform surgery. We may use them for physical rendering, as proposed in the Claytronics project at Carnegie Mellon University, that is, physical three-dimensional displays that can change shape and allow easy user interaction. In a far future, we may start to think of swarm robots as a new type of automatic construction material from which everybody can create the artifacts that surround us today. Maybe all you need to do is to obtain a seed (a module carrying a DNA string if you will) from which you can automatically grow an artifact that has superior features compared to conventional materials, such as self-repair capabilities, recyclability, and adaptability.

Let us elaborate a little on this science fiction scenario. We rely on machines and robots to aid us in our everyday life. Some machines are versatile, but only to a limited degree. Once installed, it is unlikely that machines will or can perform different types of tasks. Robots are slightly more versatile because they can be reprogrammed for new tasks, but only tasks that are within their physical limitations; e.g., a robot arm is not going to drive you to work. The problem is that the physical structure of robots, and man-made objects in general, cannot easily and in-place be changed or, using a computer analogy, be reprogrammed for a new application. For a moment imagine that this is not the case. Imagine that man-made objects can in fact change shape on demand. What if two chairs can merge and make a couch? A couch can divide and become a table and a chair? A table and a chair can melt to become a carpet? Let's look at a science fiction scenario:

John wakes up and presses the button on his bed to morph his studio apartment into a bathroom. John sighs. "Maybe it is time to update the bathroom." The bathroom is an old revision from last year with no massage chair and tiles on the walls. John calls up the catalogue from the bathroom supplier with whom he has a subscription. A miniature bathroom appears next to the sink. He flips through a couple of bathrooms until one appears that has what looks like wooden walls and even an old-fashioned toilet. The traditional toilet became obsolete years ago since the entire room is self-cleaning, but John likes the retro style. He presses "OK" and within thirty seconds his bathroom is morphed. The revised bathroom even includes his fern; he could have chosen a tropical forest theme, but he likes to have a good old-fashioned plant. Also it is a lot easier to have now that the apartment takes care of it. At that point, a signal indicates that the studio has arrived at his work place. John finishes up while the studio morphs into an office. John takes one last look in the mirror before it disappears and walks out of his studio to greet his colleagues Marvin and Louis.

In John's world everything surrounding him is made from *morphing materials* except for a few things kept for nostalgic reasons. A morphing material is a kind of material that can intelligently control its own shape. It is a material not unlike the one from which Hollywood built their Terminator robots. Applications of morphing materials are of course not limited to killing machines like the Terminator, but may literally be unlimited, depending on the characteristics of the material, including cost and energy consumption. If we invent morphing materials it may, as in John's life, completely change the way we design, manufacture, use, and recycle our everyday objects because morphing materials will provide an extreme level of versatility. Morphing materials are of course science fiction, but self-reconfigurable robots may in the long, long term be the way to implement them.

## 1.6 Structure of This Book

In order to realize the advantages of self-reconfigurable robots and perhaps in the longer term the futuristic vision described here, we face two types of challenges. One is the challenge of how to build the modules of self-reconfigurable robots; the other is the challenge of their control. These two challenges are the main topics of this book. We will look at module design in chapters 3–4 and control in chapters 5–9. The core part of the book is followed by the final chapter on research challenges in the field of self-reconfigurable robots. However, we will begin by looking at the general design goals and characteristics of self-reconfigurable robots.

## 1.7 Further Reading

We list here two articles that introduce self-reconfigurable robots, which may give the reader an alternative introduction to this field. In addition we list a doctoral thesis in which the reader can find information regarding the question of how to define self-reconfigurable robots.

M. Yim, W.-M. Shen, B. Salemi, D. Rus, M. Moll, H. Lipson, E. Klavins, and G. S. Chirikjian. Modular self-reconfigurable robot systems. *IEEE Robotics & Automation*, 14(1):43–52, 2007.

M. Yim, Y. Zhang, and D. G. Duff. Modular robots. *IEEE Spectrum*, 39(2):30–34, 2002.

E. H. Ostergaard. Distributed control of the ATRON self-reconfigurable robot. PhD thesis, Maersk McKinney Moller Institute for Production Technology, University of Southern Denmark, Odense, 2004.

tasks in many different environments. This definition again rests on the definition of what it takes for tasks and environments to be different. Here we consider two tasks or two environments to be different if in order to handle them a robot needs different behaviors or changes in its morphology that are not a natural part of its behavior (remember, self-reconfigurable robots may change configuration and thus morphology as a natural part of a behavior). In other words, tasks or environments are different if they cannot be handled by adjusting the parameters of the robot's controller or require morphological changes that are controlled by this controller (e.g., walking and galloping are different tasks whereas walking and walking fast are not). Picking up different types of objects may be different tasks if the robots need to change their behavior to do so, but often they are not different because different objects can recognized by adjusting parameters of the vision system and the gripper tool. The same is true for environments: walking on a laboratory floor is not different from walking on a road, but there is a difference between walking on a lab floor and in a sand dune (see figure 2.2). Versatility is not equally important in all applications, but must be given high priority in applications where the task may change or is unknown at the time of design. This could be the case for an extraplanetary mission or a search-and-rescue operation.

If we are to make systematic progress, it is important to be able to compare the versatility of different robots, in terms of both hardware and software. A simple way is to count and compare the number of tasks and environments that each robot can handle. However, this may be too simplistic for comparison. A fairer comparison is whether one robot is able to handle all the tasks and the environments of another robot and at least one in addition; then we can talk about one robot being more versatile than the other.

Currently, it is rare to explicitly compare the versatility of different robots because, at least in the case of self-reconfigurable robots, the potential is often there and in many cases the number of tasks that the robot can perform is limited only by how much time can be spent on programming solutions to different tasks. A typical route is therefore first to demonstrate that a robot can perform a range of basic tasks in simple environments; e.g., different styles of locomotion and manipulation in the lab. The second step is to integrate these basic tasks to solve more complex tasks or to adapt them to more complex environments. An example of a more complex, composite behavior could be to let the robot walk, crawl under a barrier, and continue to walk on the other side. An example of a simple task in a complex environment is to climb a sand dune. A final demonstration of versatility is to do complex, composite tasks in complex environments. An example may be to search a collapsed building for victims.

**Figure 2.2**
From an external perspective, rolling-track locomotion on a lab floor and in a sand dune looks similar, but we consider it to be two different tasks because the two environments are different enough to require changes in the morphology of the robot that are not a natural part of its control strategy. In this case the SuperBot had to be fitted with sand protection to successfully climb the dune [31]. (Reprinted with permission from Shen et al. [108]. Copyright 2008, American Institute of Physics)

**Figure 2.3**
An extreme example of active robustness. The ckBot breaks as a result of a hostile environment, but is able
to reassemble itself and continue its task [147]. (Courtesy of Yim, © 2007 IEEE)

are ordered or produced in small quantities may be cheap in large quantities. The cost of a prototype system may therefore be better measured in terms of the number of key components: the number of mechanical parts, actuators, and electronics components. We will return to this subject in section 2.4.1.

## 2.3 Self-Reconfigurable Robots and Conventional Robots

We have now introduced some design goals for self-reconfigurable robots. However, before we move on to describe some of the solutions, let us discuss the difference between conventional robots and self-reconfigurable robots when it comes to realizing these goals.

Conventional robots are often custom built and optimized for specific tasks (e.g., pick and place), and it is therefore difficult for self-reconfigurable robots to compete with these systems on these tasks. Rather, the power of self-reconfigurable robots is derived from their ability to perform, not only a specific task, but a range of tasks. While it may perform suboptimally compared with a custom-built robot on specific tasks, the range of tasks may be larger. Therefore, self-reconfigurable robots hold the potential to be more versatile than conventional robots because not only the control but also the morphology can be changed to perform a given task.

Redundancy and reconfigurability are key parameters in achieving active robustness. Conventional robots are often a single integrated piece of mechatronics, with little or no redundancy built in and no other way to reconfigure. This makes it difficult to recover from hardware failures. However, self-reconfigurable robots, thanks to their modular design, have a high degree of redundancy and the modules can be reconfigured. This allows self-reconfigurable robots to perform self-repair, or they may completely change configuration and behavior to maintain performance. Thus, self-reconfigurable robots have the potential to be highly robust.

Self-reconfigurable robots can adjust both their morphology and their control strategy while performing a task. This means that they can adapt to a task and an environment to a higher degree than conventional robotic systems because a fixed shape limits conventional robots. The adaptability of a self-reconfigurable robot is therefore also a crucial advantage that we should aim to exploit.

Finally, conventional robots can rely on a minimal design since they are designed for a limited range of tasks. Conversely, self-reconfigurable robots are wasteful, with many redundant mechanical and electrical components. The hope is that once designed, the modules of a self-reconfigurable robot may be mass produced, which will lower the price of modules. Therefore, self-reconfigurable robots may become cheap compared with their complexity. The design goals of self-reconfigurable robots can be summarized as follows:

just two different shapes, e.g., a mobile robot and a stationary platform. However, most metamorphic robots are designed as modular self-reconfigurable robots. These systems consist of modules that can selectively bond with other modules and have a degree of mobility to make the system self-reconfigurable. The degree to which a system is metamorphic can be measured as the range of different shapes the robot can change into. Ideally, this range is the same as the range of shapes that can be manually constructed.

### 2.4.4   Scalable

Scalability is a measure of how the performance of the robot changes with the number of modules. Increasing the number of modules can be an approach to improving several design goals. A system consisting of more modules can construct a wider range of shapes, increasing its versatility. Also, it can have a higher redundancy, which can increase its robustness. However, increasing the number of modules will, all other things being equal, also increase the price of the robot.

In many ways scalability can be compared to the complexity studies in computer science. In these studies, algorithms are rated according to their performance on large problems. In computer science, performance on small problems is not important because it is often blindingly fast. However, in self-reconfigurable robots, even small problems may take a long time because solving them involves the physical movement of modules and not just flipping of bits. So depending on the application, scalability may sometimes be sacrificed for higher responsiveness, as we will see in the next section.

Measuring scalability of course depends on the performance parameter for the task or basic functionality of interest. For example, for the basic functionality of self-reconfiguration, the question may be how the time to self-reconfigure scales with the number of modules. In locomotion, it may be how the maximum speed scales with the number of modules. In navigating a terrain, it may be how the incline that a robot can handle scales. If the project is focused on building responsive pack robots with a limited number of modules, scalability is not important. However, if the goal is to build swarm robots with myriads of modules, scalability is a crucial characteristic of the solution.

### 2.4.5   Responsive

Responsiveness is a measure of the reaction time of the robot. In locomotion, the strength of actuators is not the only limiting factor on achieving top speed. In practice, the time to detect an obstacle and take action to avoid it is more limiting. That is, the responsiveness is crucial. Responsiveness is an absolute measure of performance. In order to play tennis, a robot needs to meet certain response times, no matter how many modules it consists of. We may describe responsiveness by a reaction time and the range of configuration sizes for which this reaction time holds. Responsiveness is related to scalability. Nevertheless, in scalability studies we often accept low performance, even with only a few modules, as long as the performance scales proportionally to the number of modules. That is, it is acceptable if the time to self-reconfigure is proportional to the number of modules. Responsiveness emphasizes the point that absolute performance of a robot with a limited number of modules may also be important.

### 2.4.6  Functional

A robot design must meet the functional requirements of the application domains. Typical requirements include strength, precision, and speed in, for example, a locomotion or manipulation task. Such requirements are true for any robot, as we discussed in section 2.1, but meeting them can sometimes be difficult for self-reconfigurable robots.

Several factors contribute to this challenge in self-reconfigurable robots. To ensure versatility, these robots typically consist of only a single or a few general module types. This limits the ability to optimize for a specific task since generality is desirable. A related factor, which applies only to herd and swarm robots, is that when the number of modules is increased, the individual module's ability to directly affect the robot decreases. Therefore the modules must somehow collaborate in order for the robot to meet the functional requirements of a task. Collaboration must be supported by hardware and software and integrated into the design early.

The functionality of a self-reconfigurable robotic system is largely decided by the functionality of the modules of which it is composed. A module provides some functionality to the robot. Characteristics that are often used to describe the functionality of modules include:

*Ability to self-reconfigure*   if it can self-reconfigure in two or three dimensions

*Actuation*   number, type, speed, and strength of actuated degrees of freedom

*Sensors*   number and type of sensors

*Connectors*   number, type, speed, and strength of connectors

*Computational and communication infrastructure*   ability to communicate and perform computation

*Power*   the power source of the robot and modules' ability to share power

While these characteristics say something about the functionality of an individual module, they say little about the more important question of whether the functionality of the self-reconfigurable robot as a whole meets the requirements of a task.

## 2.5   The Use of Design Goals

The design goals and characteristics introduced in this chapter do not apply to all self-reconfigurable robots and tasks. In fact, it may be sensible to investigate how to develop a robot optimized for a specific task and thus consider only goals and characteristics relevant to that application. It may well be that as the field progresses, subtypes of self-reconfigurable robots will emerge that are optimized for specific combinations of goals and characteristics rather than all of them. We may also see that given enough time, it is in fact possible to build versatile, robust, adaptive, cheap robots. In the rest of this book we will present the means to do this in greater detail. We start by considering the mechanical and electrical designs in the following two chapters and then in the remaining chapters focus on the control aspect of self-reconfigurable robots.

## 2.6   Further Reading

Here we list a book that is not directly related to the topic of self-reconfigurable robots. However, it underlines the importance of robots having the right physical shape and being built with the right materials to be able to perform a task robustly, a lesson that is particularly important for self-reconfigurable robots because in contrast to other robots, they have a flexible shape. We also list a paper that introduces the design goals that are specific to self-reconfigurable robots.

R. Pfeifer and J. C. Bongard. *How the Body Shapes the Way We Think. A New View of Intelligence*. MIT Press, Cambridge, MA, 2006.

M. Yim, W.-M. Shen, B. Salemi, D. Rus, M. Moll, H. Lipson, E. Klavins, and G. S. Chirikjian. Modular self-reconfigurable robot systems. *IEEE Robotics & Automation*, 14(1):43–52, 2007.

# 3 Mechanical Design of Self-Reconfigurable Robots

Self-reconfigurable robots are particularly challenging to design because there is a tight coupling between hardware and software. Design problems are rarely solved in hardware or software alone, but rather solutions are found as a combination of hardware and software. For example, the complexity of writing a self-reconfiguration algorithm depends to a large degree on the motion constraints of the underlying hardware. It is therefore a design choice whether to simplify the motion constraints of the hardware or deal with these constraints at the algorithmic level. The goal of this chapter, therefore, is to enlighten the software designer about the cost of implementing a solution at the hardware level, if it is at all possible. At the same time, we aim to provide the hardware designer with a starting point for designing self-reconfigurable robots.

The software designer needs to understand the limitation imposed by the hardware in order to make a successful journey from concept, through simulation study, to hardware implementation. This chapter provides the software designer with this understanding. It is hoped that this understanding will prevent the software designer from getting stranded in simulation because unrealistic assumptions have been made about hardware.

The hardware designer often struggles to meet the demands of the software designer, who has made slightly unrealistic assumptions about the hardware. This chapter is also an aid to these hardware designers. We provide a broad overview of the solutions and design choices available. We hope this information will be useful as a starting point for designing self-reconfigurable robots and help prevent hardware designers from having to acquire some of the experience contained in this chapter at a high cost.

We start by looking at the three types of self-reconfigurable robots and continue to consider the physical geometry of modules and how to organize them in a lattice. We then describe how actuators can be exploited efficiently and discuss connector design, which is a critical aspect of any design. Finally, we briefly discuss alternative implementations of self-reconfigurable robots.

## 3.1    Types of Self-Reconfigurable Robots

All self-reconfigurable robots are slightly different because they are designed to explore different research questions within this field and represent the ideas and biases of individual mechanical designers. Nevertheless, self-reconfigurable robots are typically divided into three main types, as mentioned earlier: chain-type, lattice-type, and hybrid. Chain-type self-reconfigurable robots are particularly well suited for locomotion, but self-reconfiguration with this type of robot is often a difficult and time-consuming task. On the other hand, lattice-type self-reconfigurable robots are designed for self-reconfiguration and therefore reconfigure relatively fast and robustly, but at the expense of limited locomotive capabilities. The third type, hybrid robots, can exist in both a chain form and a lattice form and thus perform both locomotion and self-reconfiguration well. We describe these three types of robots in detail.

### 3.1.1    Chain-Type

Chain-type self-reconfigurable robots, as the name suggests, consist of chains of modules. A robot is typically formed by connecting these chains in a tree topology. For example, in the CONRO, a chain of three modules forms the spine of a hexapod robot while six one-module chains attached on the sides of the spine form the legs, as shown in figure 3.1. As shown in the figure, chain-type robots may also contain loops. A chain-type module typically has one or two rotational degrees of freedom whose axis is perpendicular to the chain. This means that longer chains have many degrees of freedom and as a consequence are flexible. The chains can be connected in many different ways, allowing these robots to perform a wide range of locomotion patterns, such as walking, side-winding, and even rolling, as shown in figure 3.1. This wide range of locomotion styles makes it possible for this type of self-reconfigurable robot to efficiently traverse any type of terrain.

Chain-type robots can also do self-reconfiguration. However, it is a difficult and time-consuming process. Self-reconfiguration consists of two steps. First, a chain bends to bring one of its end modules close to the module to which it wants to connect. Then the two modules enter a docking phase where they slowly align and approach each other and finally connect. Many self-reconfiguration steps can be combined to change the shape of the robot as a whole and thus make it possible to completely change locomotion style. A robot may start in a configuration where all modules are connected in one long chain. In this configuration the robot uses a side-winding locomotion style. Upon encountering a type of terrain where sidewinding is not effective, the robot decides to change its shape. It repeatedly bends and attaches modules along two opposing sides until it has transformed itself into a legged robot, as illustrated in figure 3.2. The robot then uses a walking gait to continue its journey. We return to this process and how it can be controlled in section 7.4 in chapter 7;

**Figure 3.1**
Chain-type robots can perform a wide range of locomotive styles. On the top is shown the CONRO self-reconfigurable robot performing a walking gait. In the middle, the CONRO robot sidewinds and performs a caterpillar-inspired, gait. (Courtesy of Stoy, © 2002 IEEE) On the bottom the PolyBot robot is rolling. (Courtesy of Yim, © 2000 IEEE)

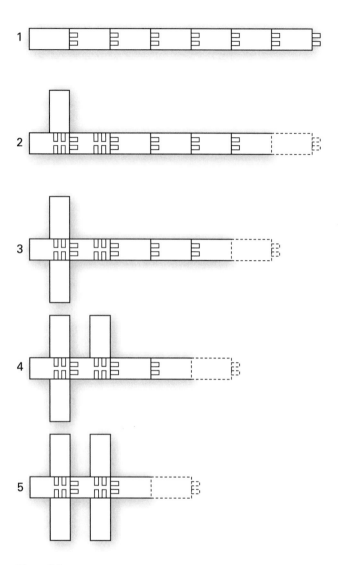

**Figure 3.2**
This is an example of how a chain-type robot can self-reconfigure from a snakelike configuration to a four-legged walker by repeatedly bending and attaching modules to its sides.

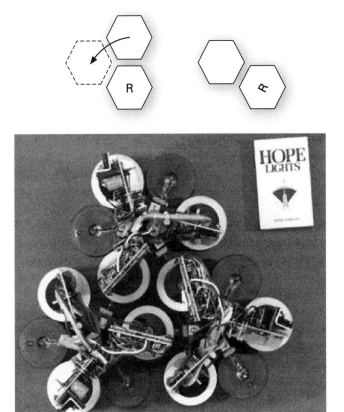

**Figure 3.3**
Some lattice-type self-reconfigurable robots move modules by rolling them around neighboring modules as shown here. The implementation differs; in some cases the module itself rolls, other times it is the neighboring module that rotates the module. On the bottom is the Fracta robot, which uses this style of self-reconfiguration. (Courtesy of Murata)

here it is just important to note that with the current implementations of chain-type robots, self-reconfiguration is a slow and slightly complicated process because it involves a search phase that requires most of the modules of a chain to act in coordination.

In summary, chain-type self-reconfigurable robots are primarily designed for fixed-shape locomotion, but through a difficult self-reconfiguration process they can change shape and thus locomotion style.

### 3.1.2  Lattice-Type

As indicated earlier, the modules of a lattice-type self-reconfigurable robot are positioned in a lattice structure like atoms in a crystalline solid. The lattice restricts the

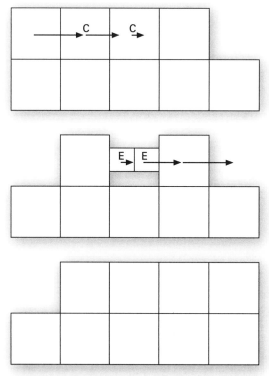

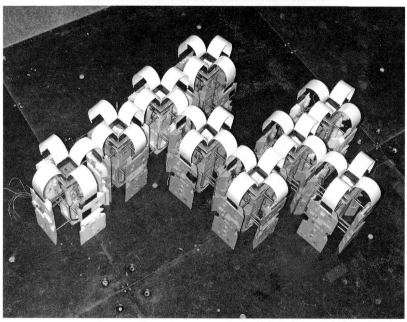

positions and orientations of modules, meaning that there is always a known, fixed distance and orientation between neighboring modules in a lattice-type robot. Modules in these robots move from one lattice position to a neighboring position in one of two ways.

Modules can move from one lattice position to another by rolling around a neighboring module (see figure 3.3) [73]. In complex three-dimensional robots of this type, a module consists of two submodules, which occupy two neighboring positions in the lattice. In these robots, movement can also be achieved by rolling around the twin submodule. In some situations this allows the module to change lattice position by itself.

Alternatively, modules that can contract and expand exploit this to move around in the lattice (see figure 3.4) [98]. In these systems, two modules contract and squeeze into the same lattice position. This contraction pulls a neighboring module with them that then can attach at a new lattice position. The contracted modules then expand and push the module on the opposite side one position further along the lattice.

Finally, modules may also slide along a surface of modules if the surface modules have a track to guide the sliding module, as shown in figure 3.5 [48]. The track idea has been implemented in two dimensions though it is difficult to expand to three dimensions because it would require the tracks to cross each other.

It is relatively easy for lattice-type robots to make a self-reconfiguration step since the lattice guides the connector alignment and thereby simplifies the mechanics and control required. In such robots, a module moving from one lattice position to another requires the help of only a few neighboring modules and sometimes a module can even move without any help. Furthermore, upon arriving at a lattice position, a module can simply assume that the neighboring modules are positioned and oriented correctly and can immediately connect to them.

Lattice-type systems, however, can perform only a limited form of locomotion through self-reconfiguration. This type of locomotion is often referred to as *cluster-flow* locomotion. In cluster flow, modules wander one by one from the rear of the robot to its front. This creates a flow of modules that causes the robot to move.

In summary, the modules of a lattice-type robot are positioned in a lattice structure that makes self-reconfiguration relatively easy. However, at the same time this structure limits the movement of modules and prevents lattice-type robots from performing flexible and efficient gaits.

**Figure 3.4**
Some lattice-type self-reconfigurable robots move modules around using the contract-and-expand sequence shown here. First, the modules labeled "C" contract and as a consequence the centers of the modules move as indicated by the arrows. Second, the modules expand again and push the neighboring module into a new lattice position, shown at the bottom. An example of a robot working this way is the Crystalline robot shown in the bottom photograph. (Courtesy of Rus)

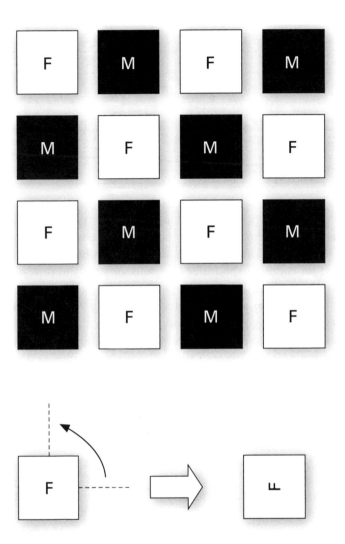

**Figure 3.7**
In bipartite robots, the modules are divided into two classes that never mix, which may be exploited to simplify connector design. In this simple example, a two-dimensional quadratic module forms the basis for a bipartite system. Each module is able to rotate 90 degrees and thus modules with female connectors (F) will be restricted to white positions and the ones with male connectors (M) to the black squares.

and white, in such a way that black positions are never next to another black position and white positions are never next to another white position. In two dimensions, a checkerboard is an obvious example. If furthermore it can be proved that once a module moves into a black position it can never move into a white position and the other way around, the system is bipartite, that is, the system is divided into two. This can be exploited to simplify connectors because instead of making unisex connectors, black modules can have female and white modules can have male connectors. The properties of the lattice and the movement capabilities of the modules guarantee that modules with the same gender never meet and therefore the reduced functionality of the connectors will never be a problem.

Although the lattice often specifies the allowed positions of modules, it does not specify the orientation. By orienting neighboring modules orthogonal to each other in the lattice, it is possible to move in three dimensions even though the individual module has only a single degree of freedom, as illustrated by the ATRON robot [82].

### 3.2.1 Submodules

Sometimes it is beneficial to cover a few lattice positions with a single module consisting of a few submodules. This does not violate the constraints of the lattice, but often it gives the individual module a higher degree of mobility. The increased mobility is obtained because submodules can rotate around submodules to move to a new lattice position. It is also possible to simplify the interface between the two submodules because a full-featured connector is not needed since the submodules are permanently connected.

A submodule-based design may also be designed as a bipartite system in which one submodule has female connectors and the other male connectors, eliminating the need for complex unisex connectors. The M-TRAN robot is an example of such a system. An alternative simplification is to utilize two types of modules, each consisting of several submodules, one module type with all female and one with all male connectors, as e.g., the Molecule robot [62] and the Roombot module shown in figure 3.8.

The submodule idea also has some downsides. The most important is that the module composed of the connected submodules can be difficult to move around because neighboring modules get in the way. This is like the difficulty involved in moving large furniture into a house. Specifically, it is difficult to move a module composed of submodules through a structure of modules and therefore it often has to travel on the surface of the structure. Furthermore, a submodule may be placed in a position where it cannot move the other submodule as desired. In both these situations, the mobility of the system is not increased; rather, the opposite is the case. These additional constraints on motion have to be overcome by clever algorithmic

Nevertheless, they are certainly not the answer to everything because they are difficult to control as well as having a relatively slow duty cycle since they have to cool down and expand before each contraction. In self-reconfigurable robots, they have mainly found their use in connectors where they are used to activate a latch or similar task.

### 3.3.6   Summary

Actuators play a crucial role in self-reconfigurable robots because they provide the robot with the ability to make chain-based motion and self-reconfigure. They also may enable an individual module to have a certain degree of movement autonomy. What we have described in this section is how the placement and orientation of actuators change the characteristics of the resulting robot. If actuators are plentiful, we have seen that we can make chains very flexible and increase the level of movement autonomy of an individual module, but they do not seem to significantly improve the ability to self-reconfigure. This should be balanced against the fact that an increased number of actuators increases the complexity and thus the cost, size, and weight of modules. If, on the other hand, the number of actuators is limited, it does not significantly change the ability to self-reconfigure, but the flexibility of chains is reduced and the individual module has no movement autonomy. This is balanced by a simplified module and thus reduced cost, size, and weight.

The actual design choice of course depends on the application. If the robot is being designed to perform chain-based motion, e.g., locomotion or manipulation, most of the time it is probably preferable to use relatively complex modules with two or three actuators. However, if the application allows less flexible chains and the emphasis is on the ability to self-reconfigure, much can be gained by using just one actuator.

Finally, we have also seen that the problem of dimensioning the actuators is not simple since the torque of the actuator influences the size and weight of the module, which again influence the torque required. In practice, it is a matter of finding a useful balance that meets the requirements of the application.

### 3.4   Connector Design

We have now looked at module geometry and lattice structures and discussed the use of actuators in self-reconfigurable robots. The last mechanical component of a module is its connectors. The connectors to a high degree decide the versatility and the ability of the resulting robot to self-reconfigure. Connectors are also important because they introduce many of the limitations of a self-reconfigurable robot. If the connector introduces few limitations, a powerful robot can be obtained. However, if the connectors are poorly designed, the whole robot suffers. Designing a connector is not a trivial challenge since—as usual in self-reconfigurable robots—there are quite a number of characteristics that it is desirable to attain.

### 3.4.1  Desirable Characteristics

We first list the desirable characteristics of connectors for self-reconfigurable robots and then discuss each in turn.

1. Small
2. Fast
3. Strong
4. Robust to wear and tear
5. High tolerance to alignment errors
6. Only use power in the transition phases
7. Transfer electrical and/or communication signals between modules
8. Genderless
9. Allows connection with different orientations
10. Disconnect from both sides

In modern three-dimensional self-reconfigurable robots, there are typically on the order of six to eight connectors in each module. This means that the size of the connector is critical because the space taken up by connectors accounts for a large fraction of the total size of a module. For instance, in the ATRON modules the connectors and their supporting structure take up about 60% of its volume. In other words, the smaller the connector, the smaller the module can be.

A self-reconfiguration step for a module consists of three substeps: disconnecting a connector, moving, and connecting a connector. This means that in order to maximize the speed of reconfiguration, it is important that the connector can perform connection and disconnection rapidly.

A connector pair connects two modules, making them one physically connected robot. The structural strength of this robot relies on the strength of the connector. If the connection is weak, it may break under stress and cause the robot to break apart. The connectors are used extensively during self-reconfiguration and thus it is important that they be robust to wear and tear.

When two self-reconfigurable robots meet, they may want to dock to become one robot (see section 7.4 in chapter 7). This requires that the two robots connect to each other through a connector. This is not a simple task because the connectors need to be aligned before a connection can be attempted. In chain mode, the alignment can be made by having a chain approach and align with a connector on another robot. This is normally based on infrared sensing and as such the process is not very precise.

Alternatively, the robots can move around to align their lattices and then make a connection. Either way, the process is not precise and it is therefore important that the connector have a high tolerance to alignment errors. Alignment errors may also occur within a single robot because a chain of modules may bend down as a result of gravity, making the tolerance to alignment errors particularly important.

In a self-reconfigurable robot, there will be many connections active at the same time, holding the robot together. If all these connections consumed power, they would quickly exhaust the energy supply and limit the lifetime of the robot. It is therefore important that connectors use power only in the transition phase, unless the robot can be powered externally because it is not required to act autonomously.

Since the connectors typically are the only physical points of contact between modules, they also have to transfer signals between modules. This may be in the form of electrical connection for power or bus communication, or it may be in the form of infrared transducers placed opposite each other for communication. It may be tempting to use wireless communication alone, but wireless does not give information about the local neighborhood of a module, such as to which module a module is connected, and therefore is rarely used alone. The connector needs to be able to transfer signals between modules. Sometimes it is made more difficult because modules can be connected with different orientations and thus the signal transfer mechanism should also handle this.

It reduces the motion constraints of a module if all connectors are unisex, meaning that any connector can connect to any other connector. The alternative is to adopt a strategy in which some connectors are male and others female (as in our normal power plugs). This is not a major problem, but it does mean that the process of connecting becomes more complicated. The modules wanting to connect have to communicate to establish whether the connectors that are about to make contact are a male-female pair. If they are not, the modules have to move with respect to each other, which may be difficult, owing to motion constraints. Another solution is to maintain a bipartite lattice structure in which males never can meet another male connector (see section 3.2). The M-TRAN robot, for instance, exploits this.

It is also sometimes desirable to connect modules with different orientations to each other. This also needs to be reflected in the design of the connector.

Finally, the last characteristic is that it is desirable if a connection can be broken from both sides. If it can, modules may not have to communicate to agree on a disconnection. Another consequence is that if a module breaks down, its neighbors are able to disconnect from it and eject it from the robot.

In summary, the connectors are crucial to the performance of the robot and as such are one of the most important design elements in a self-reconfigurable robot. In the following section we will look at possible ways to meet some of these design requirements.

### 3.4.2   Realizations of Connectors

As we can imagine after reading the previous material, designing connectors is not a trivial task given the number of desired characteristics. In fact, to date a connector has not been designed that has all the desired characteristics. Rather, the current connectors represent trade-offs among the desired characteristics.

There are three main approaches to connector design. One approach is to use magnetic connectors, another is to use mechanical connectors, and the last is to use electrostatic forces. They are all interesting approaches and choosing between magnetic or mechanical connectors mainly represents a trade-off between size and strength of the connector. Electrostatic connectors, on the other hand, are still a little way from being practically useful.

The magnetic connectors are generally smaller because they consist of permanent magnets or electromagnets, but no moving parts. They are relatively weak because they keep modules connected by using magnetic forces, which are inversely proportional to the square of the distance between magnets. The mechanical connectors, on the other hand, are strong because the retaining force is proportional to the distance of separation. However, they can be complex and take up quite a bit of space. In the following sections we will look at these three types of connectors.

**Magnetic Connectors**

The simplest possible way to make a magnetic connector is to mount permanent magnets of opposite polarity on two connection surfaces. This allows the surfaces to connect easily. However, disconnecting is more difficult because it requires the surfaces to be pulled apart with a force as strong as the one the magnetic connection provides. This force may be so strong that other connectors in the robot may disconnect instead, making it difficult to control. Therefore, some mechanism is needed to deactivate the permanent magnets in order to make it easy to pull them apart.

In the Telecube robot [122], shown in figure 3.13, the connectors are made of switching permanent magnet arrays. When the arrays of two connectors are aligned in opposite polarity, the magnets are facing each other and an attraction force is generated. The array of one side can be switched using a wire made from shape memory alloy, which contracts when it is heated by a current running through it. In this state, magnets with the same polarity face each other and the connection is broken. In one version of the M-TRAN, a similar idea [66] is used but rather than turning the polarity, a shape memory alloy is used to pull the permanent magnets apart, thus reducing the strength of the magnetic bond.

The magnetic connectors may also be used to move modules, as has been done in the Fracta [73] (figure 1.10) and Catom [58] robots (figure 3.14). In the Fracta robot, a combination of permanent magnets and electromagnets positioned around the circumference of the circular modules allows two modules to roll around each other.

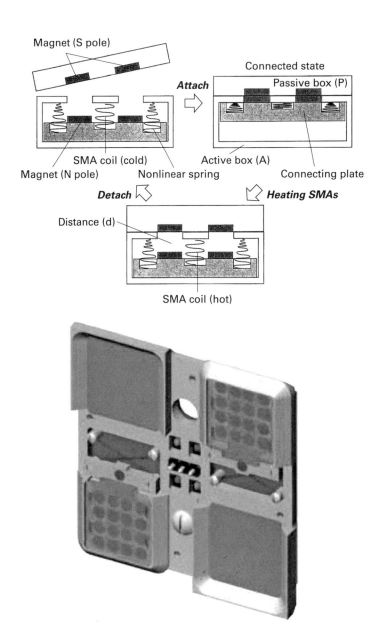

**Figure 3.13**
Top: The M-TRAN II connector. (Courtesy of Murata) Bottom: The Telecube connector. (Courtesy of Suh, © 2002 IEEE) These are examples of connector designs based on permanent magnets and shape memory alloy.

**Figure 3.14**
The Catoms use twenty-four electromagnets arranged in two rings for both connection and actuation. (Courtesy of Kirby, © 2007 IEEE)

A similar design is used in the Catoms, but in this case only electromagnets are used. The electromagnets consume power continually to stay connected, but the goal is eventually to make Catom so small that they can exploit electrostatic forces rather than magnetic forces to stay connected.

It is also worth mentioning that Velcro can in many ways serve the same purpose as permanent magnets [50] except that the ability of magnets to automatically align is lost. A common wish of the research community is a yet-to-be-invented strong, active Velcro.

The connectors that use only permanent magnets are simple and space-efficient, but, as we have mentioned, they are difficult to disconnect. The solution is to activate the magnets using shape memory alloy. This increases the complexity of the system and also increases the time to disconnect since the alloy needs to be heated before it contracts. Between these two solutions there is a trade-off between on one side simplicity and on the other the ability to disconnect. In systems where electromagnets are also used as actuators, the size of the mechanism also starts to become an issue. Finally, the strength of the connection is limited by the strength of the magnetic forces, which can be a problem in applications that require a strong connection between modules.

**Mechanical Connectors**
Mechanical connectors provide a strong connection, but at the cost of alignment problems, complexity, and size. In the earlier self-reconfigurable robots, a peg-and-latch mechanism was used for connection. In systems such as CONRO [24] (figure 3.10)

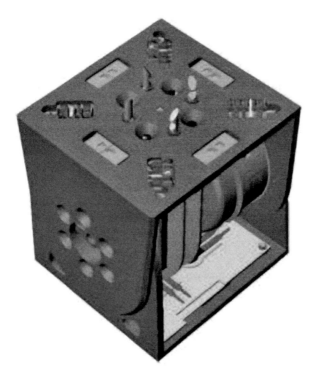

**Figure 3.15**
The third-generation PolyBot uses a mechanical connection mechanism. The connector consists of four pegs arranged in a ring around the center of two of the module's faces and matching holes in four of the module's faces. To connect, a module inserts its pegs in the holes of another module and once they are inserted, a latch closes and locks the pegs. To disconnect, a shape memory alloy actuator opens the latch and unlocks the pegs. (Courtesy of Yim, © 2002 IEEE)

and PolyBot [149] (figure 3.15), a peg of the male connector would be inserted into a matching hole in the female connector. Once inserted, a latch would fall in place and prevent the peg from being pulled out again. To disconnect, shape memory alloy would open the latch and as a result release the peg. While these mechanisms are fairly simple, it is difficult to make modules connect autonomously because of the precision needed to insert the peg in the right position and orientation.

A solution to this problem is to use a hook in the male connector that can hook onto the female connector. This mechanism was originally used in the Molecule [61] (figure 3.16) and the idea was later also utilized in the ATRON [82] (figure 1.3) and in the M-TRAN III [124] robots (figure 3.17). The hook-based design allows some correction of alignment and position errors and therefore makes forming a connection autonomously a little easier, but this type of connector is more complex and has greater space requirements. Furthermore, if the module of the male connector

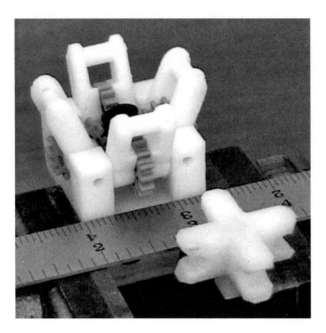

**Figure 3.16**
The Molecule pioneered the idea of having a male connector (upper left) actively grasp onto a female connector (the cross at the lower right). (Courtesy of Kotay, © 2000 IEEE)

fails, the female module cannot disconnect itself. To address this issue, the SINGO connector [102] (figure 3.18) uses two active hooks that grab onto each other. This way both sides can disconnect independently if needed, but at the cost of two actuators per connection instead of one.

The Metamorphic [86] robot and Chobie [59] form the mechanical counterpart to the Catom and Fracta robots. In these systems the connection mechanisms also provide actuation. As a result there is no disconnection and connection; the modules are always connected. The basic concept is that a peg on the male side of a module runs in grooves on the female side of another module. This allows the modules to stay connected and at the same time move with respect to each other. Although this design is not really a connection mechanism, it demonstrates that systems do not have to be built with connection mechanisms.

**Electrostatic Connectors**
A recently proposed connector type is one based on electrostatic forces [57]. In this type of connector, two opposite faces are charged with different polarity and thereby attract each other and form a strong connection. A prototype of such a system has been developed and can be seen in figure 3.19. While this type of connector has

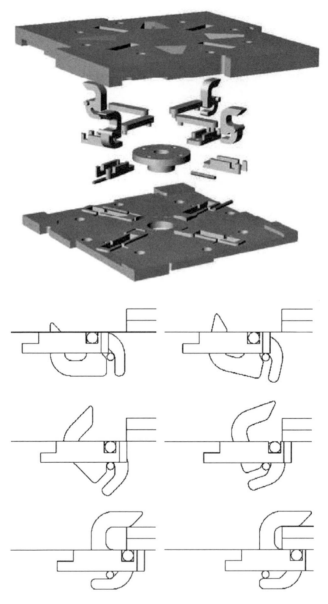

**Figure 3.17**
The M-TRAN III connector (originally developed for the robot shown in figure 3.9). The main feature of this connector is that like the Molecule's connector, it uses active hooks, but manages to do so using very little space. (Top, courtesy of Murata; bottom, courtesy of Terada, © 2008 Sage Publications)

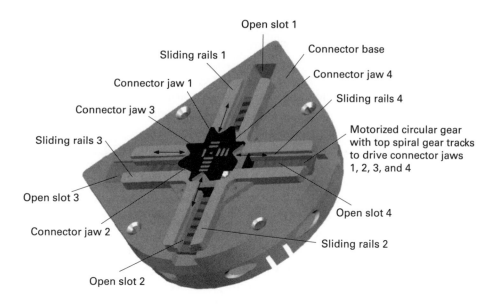

**Figure 3.18**
The main features of the SINGO connector are that it is genderless and able to disconnect from both sides. The four jaws, which are close together in the center, move in the plane of the connector face. (Courtesy of Shen, © 2008 IEEE)

many of the desirable characteristics, more research is necessary to reduce its size and increase its robustness to wear and tear.

### 3.4.3 Summary

Mechanical connectors are chosen in most modern systems because strength is considered the most important characteristic. However, this choice comes at a cost in terms of size and complexity.

Permanent magnets are also still in use because of their simplicity [147] and the availability of stronger magnets, but disconnection remains an issue when using only permanent magnets. Recently, the use of permanent magnets in combination with shape memory alloy has disappeared, probably because the alloy is slow and difficult to use. This trend is further enforced by the high level of performance of the newest generation of mechanical connectors. Electromagnets are still used because they can be used to simulate electrostatic forces [58], which may become practically useful in the near future, at least for small robots.

The advances in connector design in recent years have sparked a renewed optimism in the self-reconfigurable robot community. The newer designs of M-TRAN and ATRON show that useful mechanical connectors can be realized. These connectors have most of the desired characteristics except for being genderless and able to disconnect from both sides. The early prototype of the SINGO connector indicates that these shortcomings can be addressed as well. While being satisfied with these solutions, it is important to stay open to the opportunities of new connection technology even though this implies fundamental changes in the way we think about self-reconfigurable robots because it may allow us to realize simpler, smaller, or cheaper robots.

### 3.5 Alternative Implementations

In this chapter so far we have limited ourselves to describing some of the choices available when designing self-reconfigurable robots as defined in section 1.1. However, there are alternative ways to implement self-reconfigurable robots.

In gravity-free environments, such as under water or in space, modules can more easily move around in three dimensions. This makes it possible for modules to simply move across open space directly to the position where they are needed, rather than having to stay connected to the robot at all times. The HYDRON module is an example of an underwater module that is able to move in three dimensions [81]. In these gravity-free environments, the mechanical design problems completely change. For instance, the strength of connectors is less important and modules have to be self-propelled.

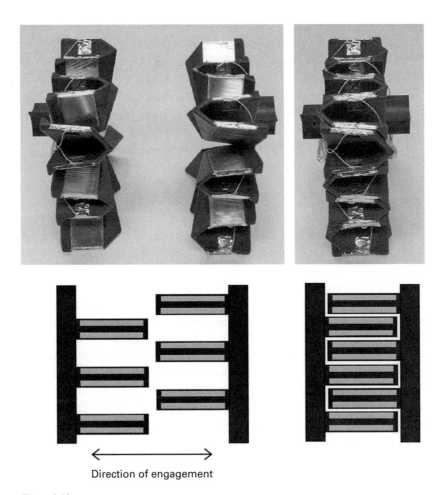

Direction of engagement

**Figure 3.19**
A connection can be made by building up a charge on two faces with opposite polarity, making them attract and form a strong connection based on electrostatic forces. Notice that the mechanical design can handle significant alignment errors and prevents disconnection in a direction perpendicular to the direction of connection. (Courtesy of Karagozler, © 2007 IEEE)

Even further removed from self-reconfigurable robots are mobile multirobot systems. These systems work only in two dimensions and this puts completely different requirements on the individual robots and their connectors. Mobile multirobot systems range in size and complexity from a few complex robots to swarms of simple robots. At the extreme range of this scale, we talk about swarm robots and although they may not share many of the mechanical design challenges of self-reconfigurable robots, they do share the problem of how to create scalable controllers for a massive number of independent modules or robots [8].

Another implementation is to use a constructor built from modules that can manipulate and connect modules to build robots in new configurations. It is of particular interest that this constructor may even build a copy of itself and achieve a limited form of self-replication [29]. This constructor-based implementation has only been investigated to a limited degree, but may hold the potential to simplify module design because only the aggregated constructor needs to provide functionality and not the individual modules.

A third option is to submerge modules in a medium that exerts forces on the modules that make them move around randomly. The medium is either water or oil, with a current providing the movement in three dimensions [137, 136] or an air-hockey table with fans providing the external forces in two dimensions [5]. When the modules meet by chance, they connect and form a larger structure and in the end form the desired configuration. This class of systems is able to perform stochastic self-assembly, whereas self-reconfigurable robots, at least in theory, are able to perform both self-assembly and self-reconfiguration. The attraction is that the modules in a system designed for stochastic self-assembly are significantly simpler because they don't need to include actuators for propulsion.

All these alternative implementations as well self-reconfigurable robots can be used to create self-replicating robots. That is, robots that are able to make copies of themselves from raw material available in the environment. For now, the raw material is limited to modules and we can therefore talk about only a limited form of self-replication. A comprehensive survey of kinematic self-replicating systems, which includes many other alternative ways to implement self-replication, can be found in *Kinematic Self-Replicating Machines* by Freitas Jr. and Merkel [53]. In the survey by Sipper [109], self-replication in software is also included.

## 3.6  Conclusion

In this chapter we have discussed some of the major design choices. We described chain-type, lattice-type, and hybrid self-reconfigurable robots and concluded that there is little reason not to design hybrid robots since it does not introduce extra complexity to make a hybrid robot compared with the more specialized chain-type and lattice-type systems. It is rather a question of picking an appropriate module geometry and lattice structure.

Another topic we have covered is the role of actuators in self-reconfigurable robots. Actuators provide modules with autonomy of movement, make chains of modules able to perform complex motions and, finally, facilitate self-reconfiguration by allowing modules to move with respect to each other. We conclude that the number of actuators is a trade-off between simplicity and movement autonomy of the individual module since chain-based motion and self-reconfiguration can be achieved using a single actuator per module.

The chapter was also devoted to discussing the design of connectors. The connectors allow modules to connect to form larger robots and as such are one of the most crucial aspects of module design. We saw that the design of connectors is a challenging task, given the number of desired characteristics. We also saw that the newest generation of connectors generally performs quite well. The choice of magnetic or mechanical connectors is largely determined by the connector strength required. Simple magnetic connectors are optimal if a limited amount of connector strength is needed and the capability to disconnect is not of great importance. For making a strong connection, mechanical connectors are best. Electrostatic forces may be an interesting alternative.

Finally, we discussed some alternative implementations of self-reconfigurable robots. In the future these implementations may provide a better way to implement these robots.

Table 3.3 contains an overview of existing self-reconfigurable robots and their characteristics. This table is complemented by table 3.4, which adds more background information. These systems and the topics we have explored here should convince the reader that there is knowledge available that makes it possible to design useful self-reconfigurable robots. The recent breakthroughs in hybrid self-reconfigurable robots and in connector design have sparked a new enthusiasm in the field. It is no longer a question of whether or not self-reconfigurable robots can be realized. This has been demonstrated in three different systems within the past few years. The question now is one of optimization and in particular optimizations for applications of self-reconfigurable robots.

Given the relative youth of the field, it is very likely that better solutions exist than the ones described here. As new technology becomes available, it is very likely that better modules can be designed. Therefore this chapter should not be seen as a recipe for how to design modules or as an attempt to maintain status quo, but rather as a starting point for exploration of new designs. Starting a module design process using the information in this chapter should make it possible to avoid the common pitfalls of self-reconfigurable robots and most likely contribute to advancing the field toward applications of these robots.

**Table 3.3**
Overview of existing self-reconfigurable robots: Mechanical design

| Robot System | Dimension | Actuator Degrees of Freedom | Connectors (Actuated) | Actuation | Attachment Method |
|---|---|---|---|---|---|
| Polypod | 3D | 0/2 | 6/2 | - / ⊥ or ‖ | Mechanical |
| CONRO | 3D | 2 | 4 (1) | ⊥ and ‖ | Mechanical, shape memory alloy |
| Polybot | 3D | 1 | 2 (2) | ⊥ | Mechanical, shape memory alloy |
| Metamorphic | 2D | 3 | 6 (3) | ‖ and ⊥ | Mechanical |
| Fractum | 2D | 0 | 3 (3) | ⊥ | Electromagnets |
| 3D-Unit | 3D | 6 | 6 (6) | ○ | Mechanical |
| Molecule | 3D | 4 | 10 (10) | ○ or ⊥ | Mechanical |
| Vertical | 2D | 2 | 1 (1) | ⊥ | Permanent magnets, mechanical |
| Telecube | 3D | 1 | 6 (6) | ‖ | Switching permanent magnets |
| Crystalline | 2D | 1 | 4 (2) | ‖ | Mechanical |
| Micro-Unit | 2D | 2 | 4 (2) | ⊥ | Mechanical |
| I-Cubes | 3D | 3 | 2 (2) | ○ and ‖ | Mechanical |
| Chobie II | $2\frac{1}{2}$D | 1 | 4 (2) | ‖ | Mechanical |
| Gear-Unit | 2D | 1 | N.A. | ⊥ | Permanent magnets |
| Catoms | 2D | 24 | 24 (24) | ⊥ | Electromagnets |
| M-TRAN II | 3D | 2 | 6 (3) | ⊥ or ○ | Permanent magnets, shape memory alloy |
| ATRON | 3D | 1 | 8 (4) | 45° | Mechanical |
| Molecube | 3D | 1 | 2 (2) | 45° | Electromagnets |
| M-TRAN III | 3D | 2 | 6 (3) | ⊥ or ○ | Mechanical |
| SuperBot | 3D | 3 | 6 (6) | ⊥ and ○ | Mechanical |
| Roombot | 3D | 3 | 10 (4) | Various angles | Mechanical |

Notes: ⊥: Perpendicular to direction of connection; ‖: parallel to direction of connection; ○: Rotation around direction of connection; N.A. = not applicable.

Overall, we have seen that although module design is indeed challenging, solutions exist to all challenges. In the following chapter we will continue to consider the design of self-reconfigurable robots, but turn our attention toward electronics.

## 3.7 Further Reading

### Types of Self-Reconfigurable Robots

We list here papers that describe a number of representative robots of each type.

**Table 3.4**
Overview of existing self-reconfigurable robots: Background information

| System | Author | Affiliation | Year | Type | Ref. |
|---|---|---|---|---|---|
| PolyPod | Yim | Stanford | 1993 | Chain | [139] |
| CONRO | Castaño, Shen, Will | USC's ISI | 1998 | Chain | [24] |
| PolyBot | Yim et al. | PARC | 1998 | Chain | [142] |
| Metamorphic | Chirikjian | JHU | 1993 | Lattice | [27] |
| Frata | Murata et al. | AIST | 1994 | Lattice | [73] |
| 3D-Unit | Murata et al. | AIST | 1998 | Lattice | [75] |
| Molecule | Kotay et al. | Dartmouth | 1998 | Lattice | [61] |
| Vertical | Hosokawa et al. | RIKEN | 1998 | Lattice | [48] |
| Telecube | Suh et al. | PARC | 1998 | Lattice | [122] |
| Crystalline | Rus and Vona | Dartmouth | 1999 | Lattice | [96] |
| Micro-Unit | Yoshida et al. | AIST | 1999 | Lattice | [150] |
| I-Cubes | Ünsal et al. | CMU | 1999 | Lattice | [129] |
| Chobie II | Inou et al. | TiTech | 2002 | Lattice | [49, 59] |
| Gear-Unit | Tokashiki et al. | Ryukyus | 2003 | Lattice | [126] |
| Catoms | Kirby et al. | CMU | 2005 | Lattice | [58] |
| M-TRAN II | Murata et al. | AIST | 1998 | Hybrid | [76] |
| ATRON | Ostergard et al. | USD | 2003 | Hybrid | [52, 82] |
| Molecube | Zykov et al. | Cornell | 2005 | Hybrid | [158, 159] |
| M-TRAN III | Kurokawa et al. | AIST | 2008 | Hybrid | [67] |
| SuperBot | Salemi et al. | USC's ISI | 2005 | Hybrid | [99] |
| Roombot | Sproewitz et al. | EPFL | 2008 | Hybrid | [110] |

Stanford, Stanford University; USC's ISI, Information Sciences Institute, University of Southern Denmark; PARC, Palo Alto Research Center; JHU, John Hopkins University; AIST, National Institute of Advanced Industrial Science and Technology; Dartmouth, Dartmouth College; RIKEN, Riken Institute of Physical and Chemical Research; CMU, Carnegie Mello University; TITech, Tokyo Institute of Technology; Ryukyus, University of the Ryukyus; USD, University of Southern Denmark; Cornell, Cornell University; EPFL, École Polytechnique Fédérale de Lausanne.

## Chain-Type

A. Castano, W.-M. Shen, and P. Will. CONRO: Towards deployable robots with inter-robot metamorphic capabilities. *Autonomous Robots*, 8(3):309–324, 2000.

M. Yim, D. G. Duff, and K. D. Roufas. PolyBot: A modular reconfigurable robot. In *Proc., IEEE Int. Conf. on Robotics and Automation*, pages 514–520, San Francisco, CA, 2000.

## Lattice-Type

S. Murata, H. Kurokawa, and S. Kokaji. Self-assembling machine. In *Proc., IEEE Int. Conf. on Robotics and Automation*, pages 441–448, San Diego, CA, 1994.

D. Rus and M. Vona. Crystalline robots: Self-reconfiguration with compressible unit modules. *Autonomous Robots*, 10(1):107–124, 2001.

**Hybrid**

S. Murata, E. Yoshida, A. Kamimura, H. Kurokawa, K. Tomita, and S. Kokaji. M-TRAN: Self-reconfigurable modular robotic system. *IEEE/ASME Transactions on Mechatronics*, 7(4):432–441, 2002.

E. H. Ostergaard, K. Kassow, R. Beck, and H. H. Lund. Design of the ATRON lattice-based self-reconfigurable robot. *Autonomous Robots*, 21(2):165–183, 2006.

**Lattice Structure and Module Geometry**

The following papers describe robots with different lattice structures or module geometry.

**2-D**

S. Murata, H. Kurokawa, and S. Kokaji. Self-assembling machine. In *Proc., IEEE Int. Conf. on Robotics and Automation*, pages 441–448, San Diego, CA, 1994.

A. Pamecha, C. Chiang, D. Stein, and G. S. Chirikjian. Design and implementation of metamorphic robots. In *Proc., ASME Design Engineering Technical Conf. and Computers in Engineering Conf.*, pages 1–10, Irvine, CA, 1996.

G. S. Chirikjian. Kinematics of a metamorphic robotic system. In *Proc., IEEE Int. Conf. on Robotics and Automation*, volume 1, pages 449–455, San Diego, CA, 1994.

B. Kirby, B. Aksak, J. Hoburg, T. Mowry, and P. Pillai. A modular robotic system using magnetic force effectors. In *Proc., IEEE/RSJ Int. Conf. on Intelligent Robots and Systems*, pages 2787–2793, San Diego, CA, 2007.

**3-D**

J. W. Suh, S. B. Homans, and M. Yim. Telecubes: Mechanical design of a module for self-reconfigurable robotics. In *Proc., IEEE Int. Conf. on Robotics and Automation*, volume 4, pages 4095–4101, Washington, DC, 2002.

S. Murata, E. Yoshida, A. Kamimura, H. Kurokawa, K. Tomita, and S. Kokaji. M-TRAN: Self-reconfigurable modular robotic system. *IEEE/ASME Transactions on Mechatronics*, 7(4):432–441, 2002.

M. Yim, J. Lamping, E. Mao, and J. G. Chase. Rhombic dodecahedron shape for self-assembling robots. Technical report, Xerox PARC, 1997. SPL TechReport P9710777.

E. H. Ostergaard, K. Kassow, R. Beck, and H. H. Lund. Design of the ATRON lattice-based self-reconfigurable robot. *Autonomous Robots*, 21(2):165–183, 2006.

## Submodules
These papers describe self-reconfigurable robot modules that consist of two submodules.

S. Murata, E. Yoshida, A. Kamimura, H. Kurokawa, K. Tomita, and S. Kokaji. M-TRAN: Self-reconfigurable modular robotic system. *IEEE/ASME Transactions on Mechatronics*, 7(4):432–441, 2002.

Keith Kotay. Self-reconfiguring robots: designs, algorithms, and applications. PhD thesis, Computer Science Department, Dartmouth College, Hanover, NH, 2003.

A. Sproewitz, M. Asadpour, A. Billard, P. Dillenbourg, and A. J. Ijspeert. Roombots—modular robots for adaptive furniture. In *Proc., Workshop on Self-Reconfigurable Robots, Systems, and Applications at IROS08*, pages 59–63, Nice, France, 2008.

## Heterogeneous Modules
Here are papers that describe examples of self-reconfigurable robots in which modules of different types exist in the same system.

C. Ünsal, H. Kiliccote, and P. K. Khosla. A modular self-reconfigurable bipartite robotic system: Implementation and motion planning. *Autonomous Robots*, 10(1):23–40, 2001.

Y. Terada and S. Murata. Automatic modular assembly system and its distribution control. *International Journal of Robotics Research*, 27:445–462, 2008.

## Connector Design

Here is a representative sample of papers that describe connectors of different types.

## Magnetic Connectors
J. W. Suh, S. B. Homans, and M. Yim. Telecubes: Mechanical design of a module for self-reconfigurable robotics. In *Proc., IEEE Int. Conf. on Robotics and Automation*, volume 4, pages 4095–4101, Washington, DC, 2002.

H. Kurokawa, A. Kamimura, E. Yoshida, K. Tomita, S. Kokaji, and S. Murata. M-TRAN II: Metamorphosis from a four-legged walker to a caterpillar. In *Proc., IEEE/RSJ Int. Conf. on Intelligent Robots and Systems*, pages 2454–2459, Las Vegas, NV, 2003.

S. Murata, H. Kurokawa, and S. Kokaji. Self-assembling machine. In *Proc., IEEE Int. Conf. on Robotics and Automation*, pages 441–448, San Diego, CA, 1994.

B. Kirby, B. Aksak, J. Hoburg, T. Mowry, and P. Pillai. A modular robotic system using magnetic force effectors. In *Proc., IEEE/RSJ Int. Conf. on Intelligent Robots and Systems*, pages 2787–2793, San Diego, CA, 2007.

**Mechanical Connectors**

A. Castano, W.-M. Shen, and P. Will. CONRO: Towards deployable robots with inter-robot metamorphic capabilities. *Autonomous Robots*, 8(3):309–324, 2000.

S. Murata, E. Yoshida, A. Kamimura, H. Kurokawa, K. Tomita, and S. Kokaji. M-TRAN: Self-reconfigurable modular robotic system. *IEEE/ASME Transactions on Mechatronics*, 7(4):432–441, 2002.

K. Kotay, D. Rus, M. Vona, and C. McGray. The self-reconfiguring robotic molecule. In *Proc., IEEE Int. Conf. on Robotics and Automation*, pages 424–431, Leuven, Belgium, 1998.

E. H. Ostergaard, K. Kassow, R. Beck, and H. H. Lund. Design of the ATRON lattice-based self-reconfigurable robot. *Autonomous Robots*, 21(2):165–183, 2006.

Y. Terada and S. Murata. Automatic assembly system for a large-scale modular structure: Hardware design of module and assembler robot. In *Proc., IEEE/RSJ Int. Conf. on Intelligent Robots and Systems*, pages 2349–2355, Sendai, Japan, 2004.

W.-M. Shen, Robert Kovac, and M. Rubenstein. SINGO: A single-end-operative and genderless connector for self-reconfiguration, self-assembly and self-healing. In *Proc., IEEE/RSJ Int. Conf. on Intelligent Robots and Systems, Workshop on Self-Reconfigurable Robots, Systems and Applications*, pages 64–67, Nice, France, 2008.

Michihiko Koseki, Kengo Minami, and Norio Inou. Cellular robots forming a mechanical structure (evaluation of structural formation and hardware design of "Chobie II"). In *Proc., 7th Int. Symposium on Distributed Autonomous Robotic Systems*, pages 131–140, Toulouse, France, 2004.

**Electrostatic Connectors**

M. E. Karagozler, J. D. Campbell, G. K. Fedder, S. C. Goldstein, M. P. Weller, and B. W. Yoon. Electrostatic latching for inter-module adhesion, power transfer, and communication in modular robots. In *Proc., IEEE/RSJ Int. Conf. on Intelligent Robots and Systems*, pages 2779–2786, San Diego, CA, 2007.

**Alternative Implementations**

Here are two papers that describe alternative ways of achieving self-reconfiguration, namely, through self-replication. However, as you may expect, the results are fairly limited.

M. Sipper. Fifty years of research on self-replication: An overview. *Artificial Life*, 4(3):237–257, 1998.

R. A. Freitas, Jr. and R. C. Merkle. *Kinematic Self-Replicating Machines*. Landes Bioscience, Georgetown, TX, 2004.

# 4 Electrical Design of Self-Reconfigurable Robots

The electronics of a module provides a computing and communication infrastructure that allows the modules to control the robot as a whole in a distributed fashion. That is, the modules form a network whose task it is to control the self-reconfigurable robot. Depending on the solutions adopted, the electronics may also need to distribute energy in the network and information obtained from sensors. The electrical design of a self-reconfigurable robot, like the design of the mechanics, is highly constrained both in terms of size and of power efficiency. In this chapter we will look at the different designs available to the electronics designer.

This chapter will provide input to a module design process. In addition, in combination with the chapter 3 on mechanical design it will form the foundation for understanding the challenges of controlling self-reconfigurable robots, which will be the topic of the remaining chapters of this book.

## 4.1 Computing and Communication Infrastructure

In self-reconfigurable robots, as is the case in so many distributed systems, it is rarely the power of the processors of the individual nodes that limits the performance of the robots, but rather the communication system. The processor for a module therefore does not need to be state of the art. The modern generation of self-reconfigurable robots uses modest processors such as the Atmel MEGA128L [52] with a 128-Kb flash memory, 4-Kb random access memory (RAM), and 4-Kb EEPROM (electrically erasable programmable read-only-memory) for permanent storage. It runs at a clock frequency of 16 MHz. Other processors are used as well, ranging from very limited processors such as the Basic Stamp 2 processor [24] to the very powerful Motorola PowerPC 555 processor [142] (see table 4.1 for a full overview). The memory of the Basic Stamp 2 may be too small for more demanding controllers, but otherwise the choice is not too important as long as it is possible to integrate it with the other subcomponents in the system and that it is relatively power efficient.

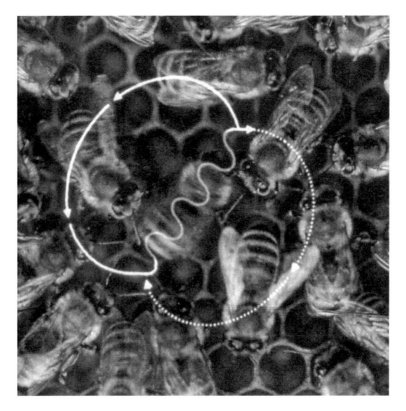

**Figure 4.1**
The famous waggle dance of bees that allows them to communicate distance and direction to a food source
[30]. (Courtesy of Chittka)

reason, as mentioned, may be to simplify the hardware, but another is that it may be
worth exploring the potential of subsymbolic communication alone and in combination with more powerful but less embodied forms of communication.

### 4.1.3 Local Communication

A step up from implicit communication in terms of complexity is local communication. The modular robotics community has realized that a module needs either
implicit or local communication to be useful. Global communication, as we will discuss later, is also useful, but only in addition to some form of local or implicit communication. This may seem surprising, but the simple reason is that modules need a
way to discover the topology of a configuration, that is, how the modules are connected. This information is needed for trivial tasks, such as asking a neighboring
module to disconnect, or more complex tasks, such as deciding which role a module

plays in a configuration. The local topology cannot be discovered using global forms of communication because messages do not contain any information about the physical position of the sender with respect to the receiver. Besides dedicated sensors, the only alternative is to program information about the module topology into the modules a priori, but this information is difficult to maintain during self-reconfiguration or when modules are added to the configuration (e.g., through a merge with another robot). In practice, it is also difficult to keep track of individual modules during experimentation.

The most common type of local communication is neighbor-to-neighbor infrared communication (e.g. [24, 98, 82]). This allows modules to detect each other before they connect as well as communicate once they are connected. It is also possible to implement local communication based on direct electrical contact through the connectors. This increases the communication bandwidth, but has the disadvantage that communication cannot be initiated before the modules are connected. This is problematic because it makes it impossible for a passive connector to ask an active connector to connect or even detect that another module is present. An infrared communication system also often provides proximity sensing and aids in aligning modules for connecting.

Local communication can simulate global communication by propagating messages in the configuration, but the propagation of messages is slower than a dedicated global communication system. This may be countered by the fact that the path the messages travel in the system may be useful because it provides information about the topology of the system. Finally, the global communication medium may become saturated when the number of modules in the robot increases.

### 4.1.4  Global Communication

Global communication allows modules to directly share information across the entire configuration through a common communication medium. While this implies that information regarding the topology of the robot cannot be obtained, as discussed in the previous section, it allows modules to communicate with a high bandwidth independently of their physical distance in the robot. This is important in applications where modules in different parts of the configuration need to cooperate tightly. For example, two arms built from chains of modules may need to cooperate to manipulate an object. It may also be the case that the data from two sensors may need to be compared. For instance, there may sometimes be a need to compare two images taken by cameras located in different parts of the configuration. Global communication is also highly useful for developing, debugging, and experimentation because it makes it easier to download new controllers, obtain the state of a specific module, or log the progress of an experiment. The problem with any global communication system is that the common medium will become saturated if enough modules use it.

One popular way to implement global communication is to use a bus that runs through the entire configuration. This provides reliable communication with a high bandwidth, but also means that the bus has to be passed across connectors, complicating their design.

An alternative is to use wireless communication. Wireless technology is not as explored as the other ways to implement communication because it only recently has become available in packages small enough to fit in self-reconfigurable robots. This is not a problem anymore, but still the research community is a little hesitant. The main problem is that few wireless communication systems of limited size and with limited power consumption have demonstrated the scalability needed to run hundreds or even tens of communicating modules simultaneously. This point may be proven wrong, owing to the fast-paced progress of wireless technology, but until that happens, wires are a more reliable alternative.

### 4.1.5 Multimode Communication

As the previous discussion indicated, local and global communication complement each other well and it is not surprising that many systems use a combination. A combination of infrared-based local communication and bus-based global communication is a popular solution [77, 142].

Another multimode approach to communication is *hybrid* communication [46]. In this approach, modules initially can perform wired communication only with neighboring modules. The modules, however, have an internal switch that allows them to connect two neighboring modules to the same physical bus and thereby make them able to communicate directly. This means that the structure of the communication bus can be controlled from software (as shown in figure 4.2). At one extreme all

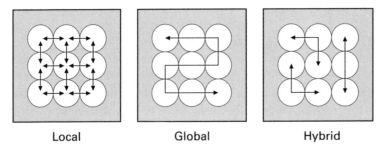

Local                        Global                        Hybrid

**Figure 4.2**
This figure explains the flexibility of a hybrid communication system. Using the same fundamental hardware local busses, a global bus or hybrid busses can be created dynamically on-line. (Courtesy of Garcia, © 2007 IEEE)

modules can communicate only with their neighbors; at the other extreme, all modules are on the same global bus, and in between, the system is split up into several hybrid busses. The downside to this approach to communication is the increased complexity of the electronics.

### 4.1.6 Summary

In summary, a self-reconfigurable robot needs local communication to discover its configuration. This can be achieved either implicitly by detecting the presence of neighbors or explicitly by local communication. Local communication also allows the propagation of configuration information. Once the configuration has been discovered, the robot may take advantage of the increased speed of global communication, which can be implemented using either a bus-based approach or a wireless solution.

### 4.2 Energy

In terms of energy, there are currently two options: on-board batteries or an external power supply. Which option is the best depends on the application. If the robot is supposed to explore unknown environments, it needs its own energy source. However, if the robot is more stationary, it is often simpler to have an external power supply. This is partly because modules can then draw as much power as they need, but also because they become lighter and easier to design if they do not have to contain batteries and power control circuitry.

It is possible to power modules from an external source using wires running to each module, but this is impractical, especially if the modules are moving. Another approach is to power the modules through the surface on which they rest, as demonstrated in the Fracta and Catoms (see figure 4.3) [73, 22].

Beyond these basic considerations, it may also be interesting to consider if modules are able to share power. This is a useful feature because from a practical point of view, it makes it possible to connect only a few modules to a charger and then through them charge the entire configuration. This avoids tedious disassembly and, after recharging is complete, assembly of the structure. In theory, it may also allow a subset of the modules of a robot to return to the charger while the remaining modules continue performing their task, and then return to charge the remaining modules. It may also allow modules that are passive and thus use limited power to share their energy with more active and power-hungry modules. The ATRON robot was one of the first to investigate how this can be implemented [52], but the feature was later removed because of short-circuiting problems. Simulations have also been used to study how to power up a structure from an external power supply [22].

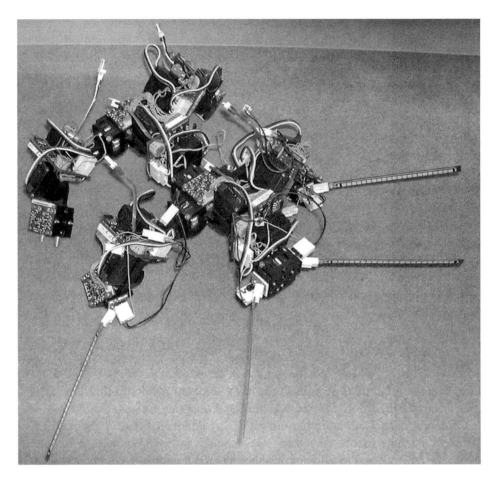

**Figure 4.4**
Like most self-reconfigurable robots, the CONRO robot was not born with sensors to sense the external environment. Therefore it was retrofitted with flex sensors to work as whiskers.

Finally, we mentioned some of the most important sensors employed in self-reconfigurable robots, such as infrared proximity sensors, accelerometers, torque sensors, tilt sensors, and cameras. Furthermore, we discussed the problem of the limited sensing capabilities of existing robots.

In chapters 2 and 3 we provided an overview of the design elements of a self-reconfigurable robot and described how they can be tied together into modern self-reconfigurable robots. This discussion also provides a foundation for understanding the challenges involved in controlling self-reconfigurable robots, which is the topic of the following chapters.

## 4.5  Further Reading

Key articles are listed under the headings of their respective sections. See the references for all articles cited in the main text.

**Computing and Communication Infrastructure**

M. W. Jorgensen, E. H. Ostergaard, and H. H. Lund. Modular ATRON: Modules for a self-reconfigurable robot. In *Proc., IEEE/RSJ Int. Conf. on Robots and Systems*, pages 2068–2073, Sendai, Japan, 2004.

A. Castano, W.-M. Shen, and P. Will. CONRO: Towards deployable robots with inter-robot metamorphic capabilities. *Autonomous Robots*, 8(3):309–324, 2000.

M. Yim, D. G. Duff, and K. D. Roufas. PolyBot: A modular reconfigurable robot. In *Proc., IEEE Int. Conf. on Robotics and Automation*, pages 514–520, San Francisco, CA, 2000.

**Subsymbolic Communication**

R. Beckers, O. E. Holland, and J. L. Deneubourg. From action to global task: Stigmergy and collective robotics. In *Proc., Artificial Life 4*, pages 181–189, Cambridge, MA, 1994.

**Local Communication**

A. Castano, W.-M. Shen, and P. Will. CONRO: Towards deployable robots with inter-robot metamorphic capabilities. *Autonomous Robots*, 8(3):309–324, 2000.

D. Rus and M. Vona. Crystalline robots: Self-reconfiguration with compressible unit modules. *Autonomous Robots*, 10(1):107–124, 2001.

**Multimode Communication**

R. F. M. Garcia, D. J. Christensen, K. Stoy, and A. Lyder. Hybrid approach: A self-reconfigurable communication network for modular robots. In *Proc., First Int. Conf. on Robots and Communication*, Athens, Greece, 2007.

S. Murata, E. Yoshida, A. Kamimura, H. Kurokawa, K. Tomita, and S. Kokaji. M-TRAN: Self-reconfigurable modular robotic system. *IEEE/ASME Transactions on Mechatronics*, 7(4):432–441, 2002.

M. Yim, D. G. Duff, and K. D. Roufas. PolyBot: A modular reconfigurable robot. In *Proc., IEEE Int. Conf. on Robotics and Automation*, pages 514–520, San Francisco, CA, 2000.

**Energy**

S. Murata, H. Kurokawa, and S. Kokaji. Self-assembling machine. In *Proc., IEEE Int. Conf. on Robotics and Automation*, pages 441–448, San Diego, CA, 1994.

J. Campbell, P. Pillai, and S. C. Goldstein. The robot is the tether: Active, adaptive power routing for modular robots with unary inter-robot connectors. In *Proc., IEEE/ RSJ Int. Conf. on Intelligent Robots and Systems*, pages 4108–4115, Edmonton, Alberta, Canada, 2005.

**Sensors**

K. Stoy, W.-M. Shen, and P. Will. On the use of sensors in self-reconfigurable robots. In *Proc., 7th Int. Conf. on the Simulation of Adaptive Behavior*, pages 48–57, Edinburgh, UK, 2002.

H. Kurokawa, K. Tomita, A. Kamimura, S. Murata, Y. Terada, and S. Kokaji. Distributed metamorphosis control of a modular robotic system M-TRAN. In *Proc., 7th Int. Symp. on Distributed Autonomous Robotic Systems*, pages 115–124, Minneapolis/St. Paul, MN, 2006.

M. Yim, B. Shirmohammadi, J. Sastra, M. Park, M. Dugan, and C. J. Taylor. Towards robotic self-reassembly after explosion. In *Proc., IEEE/RSJ Int. Conf. on Intelligent Robots and Systems*, pages 2767–2772, San Diego, CA, 2007.

# 5 The Self-Reconfiguration Problem

In the previous chapters we looked at the hardware challenges involved in designing a robot that is able to self-reconfigure. Designing such a robot was a significant problem for the research community, but the modern generation of self-reconfigurable robots has clearly demonstrated that this is not the case any more. The hardware problem has changed from being a question of how to build hardware that can self-reconfigure at all to a question of how to optimize and improve existing solutions.

In this chapter we start to look at the self-reconfiguration problem from the software side. Given hardware that has the capability to self-reconfigure, how do we actually write software that can make it happen? In section 5.1 we start to address this question by looking at three different ways of formulating the self-reconfiguration problem. This problem may seem simple but in fact has turned out to be more difficult than perhaps could be expected. In section 5.2, therefore, we continue our exploration of the control of self-reconfiguration by trying to understand the problem in detail.

The problem of controlling self-reconfiguration in fact does not have a general solution. This does not mean that the problem is unsolved, but it does mean that the solutions are hardware and application specific. The hardware and the application often allow the self-reconfiguration problem to be simplified, as we will describe in section 5.3. This simplified self-reconfiguration problem can then be solved using one of the control methods described in the following chapters. Let us start with a closer look at the problem of controlling self-reconfiguration.

## 5.1 Formulating the Problem

The idea of self-reconfiguration is that the modules of a self-reconfigurable robot are able to move around with respect to each other to change the overall shape of the robot. The problem then is one of how to move modules around to facilitate a useful change of shape. The formulation of the problem is crucial because it dictates the kind of solutions we will find. The self-reconfiguration problem has three main formulations. It can be formulated as a search problem, a control problem, or as a side effect of the robot pursuing its task:

**Definition 1 (Self-reconfiguration as search)**   Given an initial configuration and a goal configuration, find a sequence of module moves that will reconfigure the robot from the initial configuration to the goal configuration.

**Definition 2 (Self-reconfiguration as control)**   Develop a controller that makes the individual modules move in such a way that the self-reconfigurable robot as a whole will change from a given initial configuration to a given goal configuration.

**Definition 3 (Task-driven self-reconfiguration)**   Develop a controller that makes modules move in such a way that the robot as a whole performs its task. The configurations will then emerge as a secondary effect.

The search and control formulations are as old as the field itself. The first self-reconfigurable system was closer to what we today would refer to as a multirobot system. For this kind of system, it was natural to use the control formulation because the focus was on the control of individual robots or modules. The coupling between modules was not tight because modules had a high degree of mobility and therefore controlling their rearrangement was not as challenging. However, as soon as modules started to be permanently physically connected and therefore tightly coupled, it was found that the problem of rearranging modules was a computationally difficult one. It was therefore necessary to abstract some of the other aspects of the self-reconfiguration problem and focus on the search formulation. The hope was also that by looking at the problem in this more pure form, more general solutions could be found.

Unfortunately, the general solution did not materialize. In addition, while the search formulation made it easier to apply search algorithms, it also made it harder to implement the solutions found in the distributed system of modules since those solutions were developed using global information not directly available to the distributed controllers.

Today, there is a shift back toward the control formulation since solutions, even though they are not very general, are directly implementable. This is also a shift from the more theoretical search formulation to the more practical control formulation. In both these formulations the self-reconfiguration problem is decoupled from specific tasks. Tasks are specified through goal configurations. Whether this is a good way to specify tasks is questionable. It may be acceptable for simple applications such as locomotion using a specific gait or where the form itself performs a function, but in more complex tasks where the shape of the robot has to adapt to changes in the environment, it is not clear how to obtain a goal configuration.

self-reconfiguration problem is not solved. Instead, simplified versions of the problem are solved. It may seem unsatisfying to simplify the self-reconfiguration problem because as a consequence we often cannot find the optimal solution, i.e., the shortest sequence of moves. However, in practice it is enough to find acceptable, useful solutions rather than optimal solutions. In the following sections we will look at these simplifications in turn.

### 5.3.1   Meta-Modules

One of the complications of the self-reconfiguration problem is the motion constraints of the modules. These constraints turn simple transformations of shape into complex sequences of moves, if the transformation of shape is possible at all. This complication is rather frustrating since solutions exist to the self-reconfiguration problem for modules with few motion constraints. Unfortunately, it has turned out to be difficult to build such modules. Therefore, while we wait for a module design that can reduce the motion constraints, we have to find a another way. One way to do this is to use *meta-modules* [96].

A meta-module consists of a small number of modules that work together as one module. This means that rather than having the degrees of freedom of one module, a meta-module has the degrees of freedom of its constituent modules. These extra degrees of freedom can be used to reduce motion constraints and thus simplify the self-reconfiguration problem.

This simplification also comes at a cost even if it is a relatively small one: the use of meta-modules reduces the granularity of the robot [60]. This means either more modules are needed to approximate a desired shape or the approximation of the desired shape is limited by the size of the meta-module. One way to reduce this problem is to have meta-modules arise from the configuration when they are needed and have them melt back into the configuration once they are used [36]. This way, modules can be part of different meta-modules during their lifetime and thus increase the granularity of the robot.

### 5.3.2   Uniform Configuration

A module may encounter a large number of different neighborhood configurations during its movements through a lattice. It is difficult for the designer to plan or pre-program transitions in this space of neighborhood configurations because it is large. To get a feeling for the size of it, let us consider a module located in a lattice of modules. Other modules are connected to this module through a connector. These directly connected modules are neighbors. We refer to them as the 1-connection neighbors or 1-connection neighborhood. We can also consider the modules connected to the module using two connections or fewer. We refer to these as the 2-connection neighbors. In general, the *n*-connection neighbors of a module are the modules that are connected to the module through a maximum of $n$ connections along the shortest connection path.

Let us now look at the neighborhood in the case of the ATRON module. Remember that the ATRON module cannot move by itself; neighbors have to do that. In fact, sometimes it is necessary that a module rotate two modules at a time. The implication of this is that ATRON modules depend on their 2-connection neighborhood for movement. Moving an ATRON in this local neighborhood implies taking into account all the possible 2-connection neighborhood configurations. This is not a trivial task. If we look only at the different number of possible configurations of a 1-connection neighborhood, we find for the ATRON module that there are 255 different configurations. We find the number of configurations $c$ by summing the binomial coefficients of placing from one to eight modules in the eight possible neighboring lattice positions (remember that an ATRON module has eight connectors). That is

$$c = \sum_{k=1}^{8} \frac{n!}{k!(8-k)!} = 255. \tag{5.1}$$

If we extend this to the 2-connection neighborhood, the number of lattice positions increases to 38 and the number of configurations increases to $274 \cdot 10^{11}$. This number can be significantly decreased if symmetries are removed, but it still represents a challenge for search algorithms and a major problem for the programmer who has to anticipate all these configurations.

One approach to reducing this vast configuration space is to enforce a uniform structure on the allowed neighborhood configuration and then to plan or program transitions between this limited number of allowed configurations. This idea is illustrated in figure 5.6. By limiting the allowed neighborhood configurations, the configuration as a whole, which is built from this limited set of neighborhood configurations, becomes uniform [128, 85] and regular. The uniform configuration simplification is often combined with meta-modules to arrive at a limited neighborhood configuration space and a meta-module with few motion constraints that can easily move from one neighborhood configuration to another. Let us return to an example to look at the kind of reduction in neighborhood configuration space that this simplification can give. In figure 5.7 the ATRON robots are limited to the set of configurations that can be constructed by simple 2-by-2 two-dimensional meta-modules. As opposed to the individual ATRON modules, this meta-module is able to move on its own, making it depend only on its 1-connection neighborhood for avoiding collisions. So if we again consider the number of possible configurations a controller should be able to handle, this depends only on the eight direct neighbors, giving us 255 combinations; removing symmetric configurations further reduces this number to 32 [10]. The disadvantage of using a uniform structure is again one of granularity and the

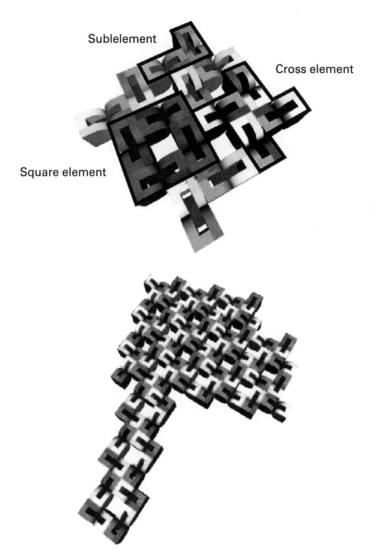

**Figure 5.6**
A uniform lattice structure of M-TRAN modules that simplifies self-reconfiguration. (Courtesy of Ostergaard, © 2004 Springer)

**Figure 5.7**
An example of combining the use of meta-modules and a uniform configuration. Top: Picture of a single meta-module for the ATRON robot. In this particular design, the meta-module consists of four ATRON modules placed in a square. Bottom: Screen shot from the simulator showing the regular grid structure of a configuration built from these meta-modules. This meta-module simplifies the motion constraints of the ATRON robot to that of sliding boxes in two dimensions. (Courtesy of Brandt, © 2007 IEEE)

fact that we can only build configurations that consist of the allowed neighborhood configurations.

### 5.3.3 Scaffolding

In scaffolding, the idea is that modules are divided into two classes: wandering modules and scaffold modules [111, 115]. The scaffold modules provide a uniform configuration in which the wandering modules can move around easily (see figure 5.8). Wandering modules move and can extend the scaffold by becoming scaffold modules in certain positions defined by the uniform configuration. The difference between a uniform configuration as presented in the previous section and a scaffold is that in a uniform configuration everything can move, whereas in the scaffold only wandering modules move. Furthermore, the scaffold is a class of uniform configurations through which wandering modules can travel.

The scaffold is grown from one initial module and therefore all scaffold modules are connected; thus wandering modules connected to the scaffold are also connected. In this way the connectivity problem discussed in section 5.2.4 can be solved locally. Furthermore, the scaffold is porous, allowing wandering modules to travel through it and avoiding the complications of hollow and solid configurations. Finally, the scaffold, being a uniform configuration, also reduces the allowed neighbor configurations. It is also possible to build a scaffold from meta-modules [78] and thus significantly reduce the motion constraints of the robot. Under all circumstances a scaffold increases the granularity of configurations.

### 5.3.4 Controller-Limited Set of Configurations

In general, we find a solution by finding a balance between the capabilities of the controllers and simplifications of the self-reconfiguration problem. For the sake of simplicity we have looked at simplifications as if they were applied before turning our attention to the controller. However, it can be an advantage to do it the other way around: take a controller and see which configurations can safely be reconfigured into and out of it and then do reconfigurations only within this set of configurations. For example, if the controller cannot handle hollow subconfigurations, then these configurations are not valid goal configurations. In a following iteration the controller can then be improved to allow a larger set of configurations [134, 135]. This means we get a simplification simply by letting the controller limit the set of configurations. The approach seems promising, but has been applied only to two-dimensional self-reconfiguration problems.

### 5.3.5 Modules Cannot Move in Parallel

The self-reconfiguration problem is often defined as one of finding a sequence of moves that will reconfigure the robot from an initial configuration to a goal

**Controller-Limited Set of Configurations**

J. Walter, J. Welch, and N. Amato. Distributed reconfiguration of metamorphic robot chains. In *Proc., Nineteenth Annual ACM SIGACT-SIGOPS Symposium on Principles of Distributed Computing (PODC'00)*, pages 171–180, Portland, OR, 2000.

J. Walter, J. Welch, and N. Amato. Concurrent metamorphosis of hexagonal robot chains into simple connected configurations. *IEEE Transactions on Robotics and Automation*, 18(6):945–956, 2002.

**Simple Intermediate Configurations**

D. Rus and M. Vona. Self-reconfiguration planning with compressible unit modules. In *Proc., IEEE Int. Conf. on Robotics and Automation*, volume 4, pages 2513–2530, Detroit, MI, 1999.

# 6 Self-Reconfiguration as Search

The preceding chapters described the mechanical and electrical design of self-reconfigurable robots and presented the self-reconfiguration problem, including possible simplifications. These pieces are parts of a solution to the self-reconfiguration problem. With the presentation of search-based solutions to the self-reconfiguration problem in this chapter, we will have all the pieces needed to form a complete solution.

When we look at self-reconfiguration as a search problem, we look at it in a very abstract form. We ignore questions related to how to implement the found solutions in a distributed system of modules, which a self-reconfigurable robot is, and how to obtain suitable goal configurations for a specific task. By ignoring these questions, it is simpler to understand the problem and the potential solutions. More important, we can more clearly understand the search aspect of the self-reconfiguration problem, which is an important aspect of any solution to this problem. This chapter assumes some basic understanding of search algorithms; it may be necessary to consult a book like *Artificial Intelligence: Structures and Strategies for Complex Problem Solving* by G. F. Luger [69] if you are unfamiliar with this topic.

In order to apply search to the self-reconfiguration problem, we need to find a suitable representation. Typically, this problem is represented as a graph in which the states are configurations and the transitions between them represent module moves. The self-reconfiguration problem is then reduced to one of finding a path in this graph from the start configuration to the goal configuration. Optimally, the solution found should be the shortest path. This path then represents the sequence of moves needed to self-reconfigure the robot from the start to the goal configuration.

Before we apply a search algorithm, it is important to look at the characteristics of this graph because it gives us an insight into the difficulty of the search problem. Maybe not surprisingly it turns out that states have a high out-degree or branching factor; that is, configurations tend to have many neighboring configurations. Furthermore, the distance, measured in number of moves, between the start and goal configurations is for all but the smallest reconfigurations fairly long. These two facts

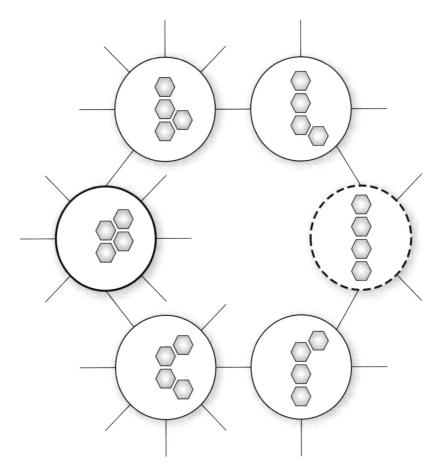

**Figure 6.2**
This is a part of a graph representing the configuration space of a self-reconfigurable robot consisting of four hexagonal modules. Each node in the graph represents a configuration and each link represents a roll either left or right of a module. Note that the unterminated links lead to parts of the graph that have not been drawn here.

third module being added will now have 10 connectors to potentially connect to, and each connection can be made by three different connectors and in four different orientations giving us 8640 combinations of three modules. In general, the number of ways to connect modules grows exponentially, and this is desirable when not talking about control because the goal of self-reconfigurable robotics is to build highly versatile robots. Because of motion constraints, the robot may not be able to self-reconfigure among all these states, but at least it gives an indication of how the size of the search space grows with the number of modules.

Now that we have demonstrated that we are dealing with a very large search space, let us consider the properties of this space in more detail. The branching factor is especially interesting. The branching factor is equivalent to the number of different moves that the modules in a cluster can perform. In general, this will be proportional to the number of modules in the cluster because every module in the cluster can potentially perform an action. If we only consider the modules on the surface of the cluster as movable, the number of possible moves will grow proportionally to the area of the surface of a sphere, which can be approximated by $n^{2/3}$ where $n$ is the number of modules. A sphere configuration is the best-case scenario because it has the lowest surface-to-volume ratio of any shape. A worst-case scenario is a chainlike configuration in which the number of movable modules will grow linearly with the number of modules. This seems like a nice slow growth, but the fact that the branching factor grows with the number of modules is troublesome since we really want to be able to control many modules and thus the branching factor will easily be on the order of hundreds or even thousands. This number can be compared to the average branching factor of chess of thirty-five.

The last property of the search we need to consider is how many moves we need in order to perform the self-reconfiguration sequence we are looking for. This number is, of course, very dependent upon how different the goal configuration is from the initial configuration, and also how capable the considered modules are. However, experiments indicate that a linear growth in the number of moves, as a function of the number of modules, is a reasonable estimate [112].

This essentially means that we have to find a path through a graph that is $n$ moves long and at every state we are confronted with $n^{2/3}$ choices of action. This spells trouble. A brute-force breath-first search would expand on the order of $(n^{2/3})^n$ states before the goal configuration was found. This means that brute-force methods are out of the question for anything but the simplest of self-reconfigurations.

## 6.3  Informed Search

The traditional way to avoid such a vast search space is to turn to informed search. In informed search, we rely on heuristics to guide us more directly through the search space from the start state to the goal state thus avoiding states that are not on the path to the goal. In this situation the vastness of the search space is of less importance because we will only explore a small, relevant part of it.

The goal of an informed search algorithm is to find a sequence of moves that reconfigure the initial configuration to the goal configuration while exploring as few configurations as possible. Usually such an algorithm is constructed from two parts. The first part is a heuristic measure of the quality of a given configuration, usually in the form of a distance metric approximating the distance from the given configuration to the goal configuration. The second part of the algorithm is a strategy for choosing which of the already explored configurations should be expanded and pursued in the next step of the search.

### 6.3.1   Distance Metrics

A heuristic distance metric may take many forms. It may involve rules of thumb that are closely linked to the motion capabilities of the modules or it may involve more complicated computations of how desirable a given state is compared with another. What we really want the metric to measure is how difficult it is to move from one configuration to another. This would reduce the search to a simple steepest desent, where the neighboring configuration minimizing the distance to the goal is chosen in each step. It is relatively simple to define the optimal metric for measuring the cost of moving from one configuration to another.

Performing a single movement $m_0$ in a given configuration $g$ is associated with a cost $C_{m_0,g} > 0$ that can represent the time or energy needed to perform the move in the given configuration. Consider a sequence of moves $m$ and an initial configuration $g$. The cost of performing all the moves in the sequence is then the sum of the costs for performing the individual moves $m_i$ in their associated intermediate configurations $g_i$:

$$C(m, g) = \sum_{i=1}^{\text{length}(m)} C_{m_i, g_i}. \tag{6.1}$$

The optimal measure for the distance between two configurations $g_1$, $g_2$ is then the cost of the optimal sequence of actions leading from $g_1$ to $g_2$. The optimal sequence is the sequence with the minimal cost:

$$\text{dist}(g_1, g_2) = \min\{C(m, g_1) : \forall m | m \times g_1 = g_2\}. \tag{6.2}$$

If no such sequences exist, then the distance is infinity.

This metric renders the general planning problem trivial to solve since it is then reduced to a simple steepest desent. The general problem is that the metric relies on the cost of the optimal action sequence leading from one configuration to another, but no efficient methods that compute this path are known. Actually, this is exactly what we need the metric to help us find in the first place.

For simple systems such as simple serial manipulators, it is possible to construct the optimal metric using the different properties of the system, but for systems with more complex motion constraints such as ATRON and M-TRAN, no one has so far succeeded in constructing an efficient method for calculating the optimal metric.

Since we cannot compute the optimal metric efficiently, it is necessary to construct some kind of approximation to the optimal metric that can be evaluated efficiently. If the metric is to offer good performance in terms of giving results close to the optimal metric, then the construction of the metric is closely related to the motion capabilities of the robot it is constructed for.

A number of metrics have been proposed for various robots. A simple example is the *overlap metric* [87, 88]. This metric is simply the number of nonoverlapping modules in the two configurations. For example, in figure 6.2 if we assume that the node to the left (bold circle) is the current state and the node to the right (dashed circle) is the goal node, then the overlap metric will give the current configuration a value of 2 since 2 is an approximation of the distance to the goal configuration. Intuitively, this makes sense. Configurations with less overlap are farther apart than configurations with more overlap. However, this simple metric represents a large underestimate because there is no way to favor one move over another unless that particular move reduces the number of nonoverlapping modules.

A more elaborate metric is the *optimal assignment metric* [88]. This metric relies on the computation of distances between individual modules in the two configurations. When the distance between all pairs of modules in the initial and goal configurations is known, an optimal assignment of modules is computed and the distance between the configurations is then the sum of the module-to-module distances. The optimal assignment can be computed in $O(n^3)$ using the Hungarian method [65]. This metric basically calculates where each module in the current configuration should end up in the final configuration, based on minimizing the sum of module-to-module distances. It then sums the number of moves each module has to make to get there, assuming that none of the other modules get in the way. However, they often do and therefore this metric is also only a significant underestimate. Furthermore, how to measure the distance between two individual modules also needs to be considered. Since the module-to-module distance should reflect the amount of work needed to move one module to the position and orientation of the other, this distance is dependent on the module morphology and possibly also the configuration. For instance, a metric for the ATRON system should have little emphasis on the rotation of the modules since it is often quite easy to rotate a module in place. For the M-TRAN system this is, however, not the case since rotations of the M-TRAN modules are inherently difficult, especially in closely packed configurations.

In order to have a good approximation for the optimal metric, it is important to consider how to construct metrics for new self-reconfigurable robots while designing them. Often it can be beneficial to rethink the module design or parts of it in order to improve the quality of the metric. In general it is desirable to have a simple relation between the actuator space and the configuration space for the system in order for the control of the system to be easy.

Let us briefly return to the complications of the self-reconfiguration problem: motion constraints, connectivity, solid configurations, hollow configurations, configurations with local minima, and overcrowding. Each of these complications negatively affects the preciseness of the estimate that the metrics provide. A metric may provide a low estimate of remaining moves when in fact most modules are trapped in a local minimum that requires many moves to escape. In this and similar situations caused by the other complications, an informed search decays to a brute-force search. This we know is intractable except in small configurations.

### 6.3.2   Search Algorithms

Let us look at some examples of implementations of informed search. In simulated annealing, transitions are made toward states closer to the goal state with a probability inversely proportional to temperature. The temperature is initially high, allowing the system to explore the search space by a random walk. The temperature is then gradually lowered to allow the system to converge toward the local maximum, which it is hoped, is the global maximum and thus the goal state [87, 88]. Unfortunately, the results even with the Metamorphing robot from Johns Hopkins University, which is two-dimensional and has few motion constraints, are discouraging: the search time as a function of the number of modules is exponential. For a Pentium-90, the search takes 6 minutes for twenty-two modules and 12 minutes for twenty-five modules. Although modern computers can reduce those times drastically, the exponential nature will still make search time prohibitively long with just another ten modules or in three dimensions.

Similar results have been obtained with iterative improvements. The idea of iterative improvement is to find a suboptimal sequence of moves that leads from the initial configuration to the goal configuration. In this sequence suboptimal subsequences are identified and optimized [26]. Two methods for identifying and optimizing suboptimal sequences are proposed. The first method is called contraction. The idea is to compare the sequence of configurations that is generated by the sequence of moves and if the same configuration is encountered twice, then the moves constituting this loop in the configuration space can be removed from the move sequence. The second method is called filtering. Here the idea is to evaluate subsequences using an upper bound for the number of moves in the sequence; if a subsequence is found that is longer than stated by the upper bound, then the subsequence is replanned and, hopefully, a shorter sequence is found. The main problems are the limited tightness of the upper bound (similar to that of the metrics) and how to come up with the initial plan.

Another way, suggested by Chiang and Chirikjian [25], is to divide the reconfiguration problem into a sequence of intermediate configurations and use simulated annealing to find the reconfiguration plans between these intermediate configurations. The problem of this approach is how to identify good intermediate configurations.

These results support our suspicion that we are dealing with a difficult search problem indeed. We know that a brute-force approach is futile, but what about using a more randomized approach? This may prevent the search from getting stuck in unproductive parts of the search space. The results produced by Brandt [9] suggest that this may be the case. In this work it was demonstrated that RRT-connect outperformed the deterministic A*. However, the limitations of the RRT-connect approach are also sobering. For a system consisting of seven modules, the search consistently failed in finding a self-reconfiguration sequence between a start and a goal configuration known to be thirty moves apart. "Failed" here means that the search was canceled after 10,000,000 states had been visited, which corresponds to a computation time of half an hour on a 1.8-GHz Intel Pentium IV processor.

Using randomized search, we produce results similar to those produced by other informed search algorithms. The documentation is too weak to conclude that one approach is better than the other, but at least it is safe to say that both come to terms with the complexity and vastness of the search space of the self-reconfiguration problem.

## 6.4   A Successful Search Requires Simplifications

The previous sections indicated that searching for solutions to a self-reconfiguration problem is, if not impossible, then time-consuming if the modules number more than a few tens. This does not mean that it is completely impossible to solve the problem using search, but it does mean that the self-reconfiguration problem has to be simplified to be tractable. In section 5.3 we introduced a range of simplifications to the self-reconfiguration problem that make the problem more approachable. These simplifications have also been successfully applied in the context of search.

A wonderful demonstration of this is the hierarchical approach to search presented in a number of papers by Ünsal, Prevas, and others. They were trying to find solutions to the self-reconfiguration problems for the I-Cubes self-reconfigurable robot. In the initial papers, the complexity of the problem is described in great detail, but only a satisfactory method for small problems is found (using A*) [131, 127, 130]. In the later papers when the simplification of meta-modules is introduced, the problem is finally solved [128, 91].

Another example of using meta-modules for search is to use a four-module meta-module for the ATRON robot, which improves the motion capabilities of the modules [10]. In essence, each of the meta-modules is a small square that can move to any of the eight neighboring positions. These simple motion constraints make it possible to use search for large problems. The meta-modules are limited to two dimensions for mechanical reasons.

Yet another example is that of Kotay and Rus [60], who use a scaffold simplification to increase parallelism by allowing modules to travel through the structure and, more important in the context of search, make it possible for modules to travel inside the scaffold following precomputed trajectories.

An example of particular interest in connection with chain-based self-reconfiguration is the use of the simple intermediate configuration simplification by Yim et al. [143]. The idea is first to calculate the sequence of actions needed for reconfiguring a general configuration into an intermediate configuration in which all modules are connected in one long chain. The next step is to reconfigure from this intermediate configuration to the goal. This is done by calculating the action sequence from the goal to the intermediate configuration and then reversing the order of the actions. Their algorithm, however, does not consider possible collisions between modules while they are moving; it only considers the problem as a high-level graph transformation problem.

These examples and other contemporary results represented a breakthrough that demonstrated that with the right simplifications, the self-reconfiguration problem is quite approachable using search.

## 6.5   From Solution to Control

The result of a successful search is a sequence of module moves. The result is produced by a centralized planner and is best executed by a centralized controller. This means that the entire robot depends on a centralized controller, which now becomes a critical component of the system that cannot be replaced in case of failure.

It is therefore desirable to make a distributed version of the plan that can be locally executed by modules. In theory this is simple to do because it only requires that all modules have a copy of the plan, know their part in the plan, and stay synchronized with all the other modules to make sure that moves are made in the correct order. In practice, however, the overhead of executing the plan is significant. Furthermore, the plan is not robust to module failures during self-reconfiguration.

It therefore became interesting to find distributed search-based solutions to the self-reconfiguration problem running on the modules themselves. This avoided the headache of distributing an essentially centralized plan and also made it possible for the system to detect and recover from module failures or obstacles in the environment.

## 6.6 On-Line Distributed Search

The search problem can be solved on-line on the modules themselves. The Pacman algorithm proposed by Butler et al. [12] is a good example. Their robot is the Crystalline robot, which is a two-dimensional robot with the capability to squeeze two modules into the same lattice position. These two features simplify the self-reconfiguration enough to be solved using a distributed approach. The algorithm works in two phases. In the first phase, virtual "pellets" are distributed in the robot, identifying paths for the modules to follow. In the second phase, modules follow the "pellet" paths to reach the final configuration. This idea is further developed and analyzed in Butler and Rus [18].

Vassilvitskii et al. [132] investigated a similar approach using the Telecubes. In this robot, an eight-module meta-module is introduced to simplify the problem. A variation of the Pacman algorithm is then used to plan a path through the structure that is converted into a sequence of basic primitives.

Instead of using on-line search to construct large-scale plans for how to move the modules, another approach is to use on-line search for solving small local reconfiguration problems. This approach has been demonstrated for meta-modules in the ATRON robot where each meta-module repeatedly calculates a small graph of the reachable positions in its neighborhood and then applies A* search to find the shortest path to the most desirable position in the neighborhood [32, 33].

## 6.7 From Impossible to Simple

In this chapter we approached the self-reconfiguration problem with search techniques. In the beginning everybody was struck by the complexity of the problem, and the shortcomings of traditional search techniques were obvious. It did not really matter whether simple search or informed search were used. Informed search was made difficult because the heuristics were poor and even the elaborate randomized search algorithms could not solve large problems because of the complexity of the search space.

The solution came when researchers started to simplify the self-reconfiguration problem. The simplifications did not reduce the usefulness of self-reconfigurable robots, but made the search problem easier. In fact, with these simplifications it turned out to be so easy that any advanced search method would do the job.

Furthermore, with the right simplifications it became possible to solve the problem directly by moving modules rather than going through a search phase first. This is the topic of the next chapter.

## 6.8   Further Reading

Key articles are listed under the headings of their respective text sections. See the references for all articles cited in the main text.

### Configuration Representation

G. F. Luger. *Artificial Intelligence: Structures and Strategies for Complex Problem Solving*. 6th ed. Addison-Wesley, Reading, MA, 2009.

G. S. Chirikjian. Kinematics of a metamorphic robotic system. In *Proc., IEEE Int. Conf. on Robotics and Automation*, volume 1, pages 449–455, San Diego, CA, 1994.

### Informed Search

### Distance Metrics

A. Pamecha, I. Ebert-Uphoff, and G. S. Chirikjian. Useful metrics for modular robot motion planning. *IEEE Transactions on Robotics and Automation*, 13(4):531–545, 1997.

### Search Algorithms

G. Chirikjian, A. Pamecha, and I. Ebert-Uphoff. Evaluating efficiency of self-reconfiguration in a class of modular robots. *Robotics Systems*, 13:317–338, 1996.

C.-J. Chiang and G. S. Chirikjian. Modular robot motion planning using similarity metrics. *Autonomous Robots*, 10(1):91–106, 2001.

D. Brandt. Comparison of A* and RRT-connect motion planning techniques for self-reconfiguration planning. In *Proc., IEEE/RSJ Int. Conf. on Intelligent Robots and Systems*, pages 892–897, Beijing, China, 2006.

### A Successful Search Requires Simplification

C. Ünsal, H. Kiliccote, M. Patton, and P. K. Khosla. Motion planning for a modular self-reconfiguring robotic system. In *Proc., 5th Int. Symp. on Distributed Autonomous Robotic Systems*, pages 165–175, Knoxville, TN, 2000.

C. Ünsal and P. K. Khosla. Solutions for 3-D self-reconfiguration in a modular robotic system: Implementation and motion planning. In G. T. McKee, T. Gerard, and P. S. Schenker, editors, *Proc., SPIE Sensor Fusion and Decentralized Control in Robotic Systems III*, volume 4196, pages 388–401. SPIE, Bellingham, WA, 2000.

C. Ünsal, H. Kiliccote, and P. K. Khosla. A modular self-reconfigurable bipartite robotic system: Implementation and motion planning. *Autonomous Robots*, 10(1):23–40, 2001.

C. Ünsal and P. K. Khosla. A multi-layered planner for self-reconfiguration of a uniform group of I-Cube modules. In *Proc., IEEE/RSJ Int. Conf. on Intelligent Robots and Systems*, volume 1, pages 598–605, Maui, Hawaii, 2001.

K. C. Prevas, C. Unsal, M. O. Efe, and P. K. Khosla. A hierarchical motion planning strategy for a uniform self-reconfigurable modular robotic system. In *Proc., IEEE Int. Conf. on Robotics and Automation*, volume 1, pages 787–792, Washington, DC, 2002.

D. Brandt and D. J. Christensen. A new meta-module for controlling large sheets of ATRON modules. In *Proc., IEEE/RSJ Int. Conf. on Intelligent Robots and Systems*, pages 2375–2380, San Diego, CA, 2007.

K. Kotay and D. Rus. Algorithms for self-reconfiguring molecule motion planning. In *Proc., IEEE/RSJ Int. Conf. on Intelligent Robots and Systems*, volume 3, pages 2184–2193, Maui, Hawaii, 2000.

M. Yim, D. Goldberg, and A. Casal. Connectivity planning for closed-chain reconfiguration. In G. T. McKee, T. Gerard, and P. S. Schenker, editors, *Proc., Sensor Fusion and Decentralized Control in Robotic Systems III*, volume 4196, pages 402–412. SPIE, Bellingham, WA, 2000.

**On-Line Distributed Search**

Z. Butler, S. Byrnes, and D. Rus. Distributed motion planning for modular robots with unit-compressible modules. In *Proc., IEEE/RSJ Int. Conf. on Intelligent Robots and Systems*, volume 2, pages 790–796, Maui, Hawaii, 2001.

Z. Butler and D. Rus. Distributed motion planning for 3-D modular robots with unit-compressible modules. In *Proc., Workshop on the Algorithmic Foundations of Robotics*, pages 435–451, Nice, France, 2002.

S. Vassilvitskii, M. Yim, and J. Suh. A complete, local and parallel reconfiguration algorithm for cube-style modular robots. In *Proc., IEEE Int. Conf. on Robotics and Automation*, volume 1, pages 117–122, Washington, DC, 2002.

D. J. Christensen. Evolution of shape-changing and self-repairing control for the ATRON self-reconfigurable robot. In *Proc. of the IEEE Int. Conf. on Robotics and Automation (ICRA)*, pages 2539–2545, Orlando, FL, 2006.

D. J. Christensen. Experiments on fault-tolerant self-reconfiguration and emergent self-repair. In *Proc., Symp. on Artificial Life-part of the IEEE Symposium Series on Computational Intelligence*, pages 355–361, Honolulu, Hawaii, 2007.

# 7 Self-Reconfiguration as Control

In chapter 6 we viewed the self-reconfiguration problem as a search problem. The advantage of this view is that we look at the problem from a global perspective, which gives us easy access to global information that we can exploit to generate solutions. For example, it is easy to coordinate the actions of modules far apart in the configuration. However, the fact that solutions rely on global information also makes it difficult to transform them into local, distributed controllers. In this chapter we will attack the self-reconfiguration problem from the other end, so to speak. We will develop solutions directly in terms of local, distributed controllers and thus view the self-reconfiguration problem as a control problem. We take a local view and look at the self-reconfiguration process from the viewpoint of the individual modules. This viewpoint makes it easy to generate controllers, but makes it difficult to discover and handle reconfiguration problems that require global knowledge. In the local or control-based view, the self-reconfiguration process can be seen as a swarm of independently acting modules moving from an initial configuration to a goal configuration.

All controllers in this swarm need a movement strategy and a representation of the goal configuration. The goal representation makes it possible for a controller to decide whether or not the module that it is controlling has reached a goal position. And in case it has not, the movement strategy makes the controller able to decide where to move the module in order to move it closer to a goal position. If all modules in the robot move step-by-step closer to goal positions, the robot as a whole will end up in the goal configuration.

This does involve a little luck. The complications of the self-reconfiguration problem make it difficult to design a controller that can solve the problem in general. However, by simplifying the problem and using a trick or two, it is possible to develop useful controllers that go a long way toward solving the self-reconfiguration problem. In particular this is true if the specific application allows us to make assumptions about the goal configurations. This is the topic of chapter 8. Here we will start by looking at strategies for moving modules toward goal positions.

## 7.1    Movement Strategy

The movement strategy answers the question of how a module finds a path from its initial position to a position in the goal configuration. In the search-based approach, we generate a sequence of moves that reconfigure the robot from a start configuration to a goal configuration. Implicit in this sequence of moves is a description of the path each module has to travel to reach its goal position. These paths may with some difficulty also be generated in a distributed system. However, in general they are fragile since they rely on all modules being able to follow their path in a timely manner. This is a shaky assumption because obstacles in the environment may obstruct the paths and modules may break or be delayed. These problems can be addressed using repeated replanning and careful coordination, but if they happen too often compared with the time it takes to generate a new plan, this becomes inefficient. Therefore, it is often desirable to use a more robust and fault-tolerant approach.

In self-reconfiguration as control, the key idea is that each module only needs to know where to move next. If modules on average move toward the goal configuration one move at a time, the robot as a whole will eventually converge to the goal configuration. This approach allows problems to be handled when they occur, with limited consequences for the robot as a whole. In the following sections we will look at how a module can decide on its next move.

### 7.1.1    Randomness

The easiest way to select among possible moves is to pick a random one [73, 51]. This approach is attractive because it is guaranteed to eventually move the module to a goal position, but the downside is that it is rather inefficient in terms of number of moves and time. However, this is not always the case.

If we see modules as a swarm of particles moving around randomly, it is clear that once a goal position is opened, it will be filled quickly if the density of randomly moving, spare modules in the vicinity is high. So the trick is to have a large number of spare modules available during the entire self-reconfiguration process. Rosa et al. [92] pursued a related approach. In their work, it is holes in the configuration that move randomly around rather than modules. The life of a hole has three phases. First, the hole is created on a growth surface of the configuration. Second, the hole wanders around randomly inside the configuration. Finally, the hole is destroyed when it reaches a destruction surface of the configuration. The holes do not move in a random direction at every step; rather, they change direction when they bump into other holes or the surface of the configuration, as illustrated in figure 7.1.

This approach is mainly useful in swarm robots, but even in swarm robots extra care has to be taken to handle specific motion constraints and to prevent modules

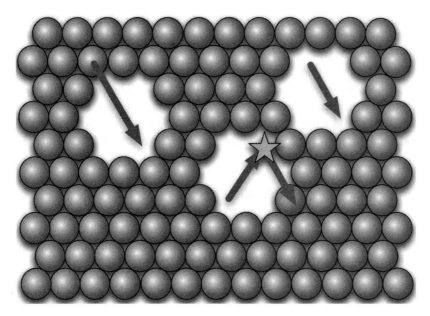

**Figure 7.1**
It is possible to focus on moving holes rather than modules. In this example holes move around randomly inside the configuration and bounce off each other, as indicated by the arrows. (Courtesy of De Rosa)

on the surface from disconnecting. In pack or herd robots where all modules are needed, it is inefficient to rely on randomness because too few spare modules are available at any given time.

### 7.1.2  Local Rules

Local rules can also control the movement of modules [16, 84]. A rule generally consists of a condition based on the configuration of the neighborhood of the module and an action. The action is performed if the neighborhood configuration matches the condition. A condition, for instance, may take the form of a bit-vector representing whether a neighboring module is connected to a connector. For example, for the ATRON module with eight connectors, the condition may be [01000010]. This condition is satisfied if neighboring modules are connected to connectors one and six. In a similar way, connector gender, orientation, etc. can be represented. The action can also be many different things, such as disconnecting a connector or moving a module. An example of a rule that disconnects connector one if it is connected could for instance be:

$$[01000010] \rightarrow \text{disconnect}(1). \tag{7.1}$$

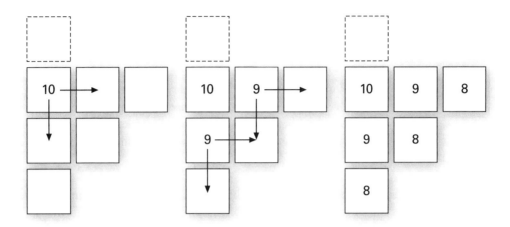

**Figure 7.3**
This figure shows how an artificial chemical diffuses through a configuration of modules (represented by boxes). First a module identifies its neighbor's position as a goal position (dashed box) and raises its concentration to 10 (left). It then communicates 10–1 to its neighbors (middle). These neighbors then communicate the highest number they have received minus 1 to their neighbors (right). This process continues until all modules know the number of connections between them and the goal. This information can then be used to calculate a gradient for the modules to follow to get to the goal position.

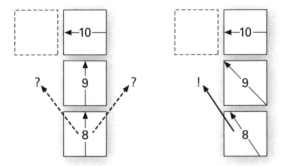

**Figure 7.4**
If the gradient calculation is based only on local information, it is not obvious on which side a module should pass another one if more options are available (left). However, with the addition of a propagated gradient vector that maintains the history of propagation, this uncertainty can be eliminated (right).

[112]. This improvement will make modules follow a more optimal path, but at the cost of an increased amount of communication.

The basic propagation algorithm outlined here is simple, but it does require quite a lot of communication. Ostergaard [80] has proposed a more complex but optimal algorithm that does not communicate more than necessary.

The advantage of gradient attractors compared with coordinate attractors is that the gradient follows the structure of the robot and therefore local minima are avoided when a module travels up the gradient. If scaffolds are used, the gradient can be propagated through the structure because scaffolds allow modules to travel through the structure. Alternatively, the gradient can be propagated along the surface of the configuration to ensure that modules can actually travel in the direction indicated by the gradient. In general, gradients can take motion constraints into account and only be propagated along movable paths. This, however, has not been explored in great detail. The disadvantage of gradients is the communication needed to maintain the gradient and the time it takes to propagate it. Also, often the motion constraints of a specific module type may complicate the action sequence needed to follow the gradient. In such cases the local planning can be utilized to allow the module to find the action sequences necessary for following a gradient, or meta-modules with simple motion constraints can be introduced [32].

### 7.1.5 Recruitment

A module that has an unfilled goal position next to it may also decide to *recruit* a spare module rather than creating a gradient for spare modules to climb. The module recruits a spare module by doing a distributed-depth first search. When a spare module has been located, it travels up the tree generated by the search until it reaches the root, where it fills the unfilled goal position. The search can take motion constraints of the robot into account and thus ensure that the spare module can actually travel up the tree [12, 20].

In comparison with recruitment, the gradient attractor can be seen as a breath-first search with a cutoff at depth $n$. This means that the gradient attractor is likely to attract the closest module, but at a cost of more communication. It is interesting to note that since the configuration is a fully distributed system, the parallel breath-first search will on average locate a module faster than a depth-first search because each branch of the tree is searched in parallel. The gradient attractor, however, has no way to ensure that only one module is attracted. This may be a disadvantage, but often it is not because if one module is needed in an area, often more modules are needed later. Imagine building a chain of modules: if we recruit modules from one end of the configuration one module at a time, the time to complete the chain is $O(n^2)$ where $n$ is the length of the chain, because on average each module has to travel $(1/2)n$ and there are $n$ modules. However, a parallel recruitment using a gradient attractor completes in time $O(n)$, which is the time it takes for the last module to reach the end of the chain.

### 7.1.6 Choosing a Movement Strategy

We have described five ways that modules can decide which way to move. Randomness is a promising option in swarm robots where parallelism can be exploited to increase the probability that a module will fill a goal position by chance. Moving randomly is the simplest possible solution and as such is suitable for minimalist modules. However, in pack or herd robots or in general configurations where there are few spare modules, randomness is a bad choice because it will be inefficient both in terms of time and of moves.

For modules with more powerful hardware for local communication, a significant increase in performance can be gained by using one of the remaining three methods. Coordinate attractors are easy to implement, but do not deal well with local minima and should therefore be used only in combination with randomness or a local search. Alternatively, gradient attractors or recruitment can be used since they deal well with this problem. The choice depends on whether it is desirable to attract one or more modules to the area of an unfilled goal positions. Often this is the case and then gradient attractors are a good option. If not, then recruitment is the better option.

Finally, it is important to note that a combination of methods is often best. Different methods can be applied in the part of the configuration where they are most suited. For example, randomness may work well inside a configuration where the density of modules is high, but on the surface where the module density is lower and where each module therefore becomes relatively more important, more care has to be taken and thus, for instance, local recruitment may be necessary.

### 7.2 Representation of the Goal Configuration

In section 7.1 we saw how modules can be controlled to move in a desired direction. However, the modules also need to know when to stop. That is, they need to be able to detect when they have reached a goal position. Knowing this means that the module either explicitly or implicitly needs to represent the goal position and its own current position. The choice of representation is important because often each module needs to store the representation.

### 7.2.1 Volume

The simplest way to represent a goal configuration is to keep a list of goal coordinates [148]. In order to use the representation, a module needs to keep track of its position in the lattice, but otherwise it is straightforward to check where the nearest

| Resolution | Low | Medium | High |
|---|---|---|---|
| Modules | 32 | 4512 | 34493 |
| Boxes | 3 | 168 | 1152 |

**Figure 7.5**
A goal shape can represented and approximated with a set of overlapping boxes as shown here in the case of a Boeing 747. The more boxes, the smaller they can be and therefore the approximation has higher resolution. (Courtesy of Stoy, © 2004 IEEE)

goal position is or whether a neighboring position is a goal position. The downside to this way of representing the goal configuration is that its size increases with the number of modules in the goal configuration.

A more space-efficient representation is to use overlapping boxes [115]. A box is represented by the position of two diagonally opposite corners. All positions within this box are then part of the goal configuration while those outside are not. To represent more complex configurations, a set of boxes is used. These boxes can overlap. In this situation, a position is a goal position if it is contained within at least one box. The size of this representation scales with the complexity of the configuration and not the number of modules (see figure 7.5). The representation can also be scaled to build the same goal configuration, but in different sizes [116].

Finally, it may also be possible to represent the goal configuration as being enclosed by a surface, similar to the way three-dimensional models are represented in computer graphics. However, the complexity of checking whether or not a position is inside may be computationally unattractive for embedded processors.

### 7.2.2 Set of Connection Types

The volume-based way of representing a configuration relies on the modules' ability to localize in three-dimensional space. Even though this is fairly simple to do, it requires that the modules have local communication. If modules do not have local communication, a way to represent the goal configuration is as a set of connection types [73]. In this representation, the local topology decides whether a position is a goal position (see figure 7.6).

The advantage of this representation is that it relies purely on local information and may be implemented on modules that do not have communication capabilities.

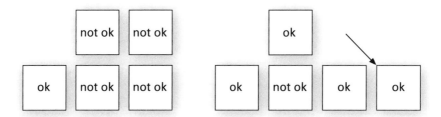

**Figure 7.6**
This figure shows how modules can detect a goal configuration using a set of connection types. Left: An example of a set of connection types for a chain configuration. Misplaced modules can detect that the goal configuration has not yet been reached. Right: A module has decided to move and the configuration is now a little closer to the goal.

An important drawback is that reaching the goal position is not guaranteed because there may be many configurations that satisfy the set of connection types. Also, as can be seen in the left part of figure 7.6, the module that can detect that the goal configuration has not been reached cannot move and thus the robot is stuck in this configuration. Finally, it is not obvious how to come up with connection sets for more complex configurations, although they have been demonstrated to be acceptable for small, simple configurations [73].

### 7.2.3   Transition Rule Sets and Growth

An approach that represents a compromise between volume-based representations and representation as a set of connection types is the transition rule set. The transition rule set uses only locally available information, just as the set-of-connection-types representation, but is as powerful in terms of representation power as the volume-based representations. The transition rule sets are not only a representation of the goal configuration, they are also a representation of how to build the goal configuration. Reconfiguration using transition rule sets can be viewed as a growth process. A module is typically assigned an identification number (ID). It then looks up its ID in the representation and finds the IDs of neighboring positions. If it finds a module in a neighboring position, its ID is communicated to this module. The neighbor then looks up this ID and propagates IDs to its neighbors and so on. If a neighboring module is not available, it can be attracted using an appropriate method. In this way the configuration grows from the initial seed module guided by a transition rule set, as shown in figure 7.7.

It is difficult to make the transition rule set by hand. Therefore we need a systematic and automatic way of transforming a humanly understandable description of a desired configuration into the rule set we will use for control. One option is to have a rule set generator which as input takes a three-dimensional model and as output pro-

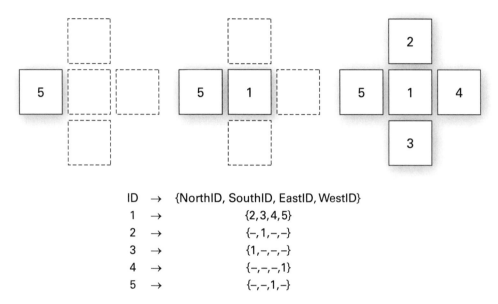

$$
\begin{array}{ccc}
\text{ID} & \to & \{\text{NorthID, SouthID, EastID, WestID}\} \\
1 & \to & \{2,3,4,5\} \\
2 & \to & \{-,1,-,-\} \\
3 & \to & \{1,-,-,-\} \\
4 & \to & \{-,-,-,1\} \\
5 & \to & \{-,-,1,-\}
\end{array}
$$

**Figure 7.7**
Transition rule sets are used to drive self-reconfiguration through growth. Bottom: A rule set is defined that using a given module ID, maps to the IDs of neighboring modules. Top: A construction sequence from left to right is shown in which an initial seed module is given an ID and from this module the desired configuration grows, guided by the rule set.

duces the corresponding rule set [111, 112]. The rule generator takes a closed three-dimensional model and a specified starting point inside the model. From this starting point the model, through a process similar to three-dimensional flood filling, is filled up with nonoverlapping boxes (or voxels) representing modules. Each box whose center is contained within the volume is given a unique ID. The algorithm then outputs a set of transition rules that specify the neighboring relationships between IDs.

During the generation phase it is also possible to create a rule set that enforces a construction sequence that builds the inner part of a configuration before it builds the outer one [51]. This avoids the complication of hollow or solid subconfigurations.

The generator in these two approaches handles the complications that can be handled only at the global level, leaving the controller with rules that can be executed using only local information. One potential problem with this approach is that the size of the rule set is proportional to the number of modules. The approaches do not rely on global coordinates, but on the ability of the modules to globally agree on directions; otherwise the rules cannot be applied. If it can be assumed that modules globally agree on direction, it is easy to generate a global coordinate system and thus the difference between this representation and the volume-based representations is small. The key difference is that the rule sets, besides representing the goal configuration, also answer the question of construction sequence, which the coordinate-based approaches do not.

### 7.2.4 Choosing a Representation

We have now seen a range of ways to represent the goal configuration. Some of the representations maintain a purely local view, such as the set-of-connection-types representation. This representation is easy to implement but does not handle the complications of the self-reconfiguration problem.

The volume-based representations, on the other hand, take a global perspective, building on the assumption that the modules share a global coordinate system, which is not an unreasonable assumption. These representations make it easy for modules to decide if the goal configuration has been reached, but do not by themselves guide the modules to reach it.

The transition rule set approach represents a combination. Some of the global complications of the self-reconfiguration problems are handled by an off-line rule set generator. The remaining complications are handled by a simplification of the self-reconfiguration problem, so that the controller can rely on purely local information.

In general, a representation is not chosen in isolation, but in combination with module, movement strategy, and simplifications of the self-reconfiguration problem.

### 7.3   Complications

Equipped with the knowledge of the previous two sections, you may feel ready to implement a controller, but that would be premature. There are still the complications of self-reconfiguration that need to be addressed. Controlling self-reconfiguration would be fairly easy if the methods described here could do the job, but this is not the case since you may also want your algorithm to be able to deal with the complications of connectivity, overcrowding, motion constraints, local minima, and hollow and solid subconfigurations.

Connectivity, as you may recall, is the problem of making sure that all the modules of a robot stay connected during self-reconfiguration. This problem has been addressed in a number of ways that have different advantages and disadvantages. It is possible to apply an algorithmic solution that before each disconnect does a local distributed search to ensure that the configuration stays connected after this particular disconnect [39]. Most of the time this search is local and is therefore relatively fast, but in the worst case it may involve all the modules of the system. However, it is probably acceptable to have a few slow self-reconfiguration steps if the rest are fast, so if the overhead of running this algorithm is acceptable, it is a useful solution.

Another option is to insist that the modules will move only when the connectivity problem can be solved locally [74]. This makes it possible to quickly check if a move causes a disconnection, but may in some cases prevent a move in a situation where it is actually possible if a larger part of the configuration is considered. This may not present a problem as long as the local configuration changes later, allowing the move.

A compromise between these two solutions is possible if scaffolds are used [115]. In this situation a search can be performed to check if the neighboring modules to a module will remain connected to the scaffold after it has moved. This may limit the search because the scaffold is never far away.

Finally, it has been proposed to use a sensor to decide whether a given disconnection would break the connectivity [145], but how to create this sensor is not obvious.

Beyond what we described in the previous sections, the remaining complications have not been addressed from a control point of view, but are solved by using a simplification of the self-reconfiguration problem. Unfortunately, even with these simplifications, controllers fall short of solving the problem in general. A good example is the solution proposed by Stoy [114]. In this solution scaffolds, modules with simplified motion constraints, and gradient attractors are used to solve the self-reconfiguration problem, but the problems of overcrowding and constraints on realistic motion remain unaddressed. Another is that of Rosa et al. [92], who, in principle, solve the self-reconfiguration problem in two dimensions, but only by assuming simple motion constraints that may not generalize to three dimensions.

The previous two examples are studies done in simulation only. For researchers starting from hardware, the problem of motion constraints is often the first encountered and the most serious one. Christensen and Stoy [36] attack this problem by using carefully designed meta-modules and local planning. This approach also to some degree manages to control self-reconfiguration, but meta-modules tend to get stuck and thus the goal configuration can only be approximated.

## 7.4  Docking and Merging

In chain-type self-reconfigurable robots, a self-reconfiguration step is not as simple as it is in lattice or hybrid robots. In these robots the lattice guarantees that if actuators are locked in specific positions, a connection with a neighboring module is almost guaranteed to be successful. In chain-based robots this is not the case. Chain-based robots need a longer chain of modules to reach a position to connect to and thus complete a self-reconfiguration. Since the chain is longer, it is more sensitive to imprecision of the positioning of the actuators because the small imprecision of each module adds up to a large imprecision in the end module of the chain.

# 8 Task-Driven Self-Reconfiguration

In chapters 6 and 7 we looked at self-reconfiguration as a problem in its own right, defined by an initial configuration and a goal configuration. This strategy eliminates all factors that seem irrelevant to the problem. We can focus on the self-reconfiguration problem without having to worry about a specific task. We may also think that a general solution arrived at this way may easily be adapted to a specific task, given that it can provide us with an initial and a goal configuration. In fact, maybe the task itself is to change into a specific configuration.

These observations are certainly valid, but there are two reasons for why we would want to look at self-reconfiguration in the context of applications beyond changing between specified configurations. The first reason is the obvious one that many applications cannot provide us with a desired goal configuration. For example, in a task involving grasping an unknown object, it seems difficult to specify a suitable configuration. In a task where the robot has to support a collapsing roof, it seems difficult and so on. In many tasks where the task environment is not known beforehand, is dynamic, or is complicated, determining a goal configuration may be intractable. The second reason is that maybe the self-reconfiguration problem is a problem simply because it is defined in terms of an initial and a goal configuration. Maybe a self-reconfiguration algorithm focused on the task-achieving process of self-reconfiguration rather than the goal is easier to design and implement. One way to look at the self-reconfiguration problem described by an initial and a goal configuration is as a worst-case scenario. It leaves the robot no freedom to choose configurations that could ease the self-reconfiguration process because, in the end, it has to squeeze itself up into the tight corner of the configuration space specified by the goal configuration. Thus, by focusing more on the process of self-reconfiguration, we may find a compromise between a suitable configuration for the robot and the task environment. In the end the specific goal configuration often is not important as long as the configuration is appropriate for the task.

In this chapter we will explore the task-achieving approach to self-reconfiguration. In the previous two chapters we discussed solutions that may be task-achieving, but certainly are driven by goal configuration. In this chapter we will, per the earlier discussion, focus on solutions to self-reconfiguration that are task driven but do not contain a representation of the goal configuration.

## 8.1   Locomotion through Self-Reconfiguration

In this section we will look at locomotion in the context of self-reconfigurable robots. According to the *Oxford American Dictionary*,

Locomotion is the ability to move from one place to another.

Locomotion is interesting to study because it is a fundamental ability needed in a wide range of applications. Furthermore, self-reconfigurable robots have an advantage when moving through an environment, because they can adapt their shape and change locomotion patterns to fit the environment. This ability can, for instance, be important in a search-and-rescue scenario. In this scenario, the robot can start in a loop configuration in order to achieve high speeds to get to the general area of interest quickly. Upon arrival, the robot can reconfigure into a snake configuration and make a careful search of the area.

Locomotion can be achieved in both dynamic and fixed configurations. In fixed-configuration locomotion, movement is achieved by controlling the joints of the modules. This topic is treated in chapter 9. In dynamic configurations, locomotion is achieved through self-reconfiguration. This type of locomotion is often referred to as *cluster-flow* or *water-flow* locomotion (see figure 8.1). The idea is that modules from the back of the robot move toward the front of the robot. This process is repeated and locomotion is generated. This mode of locomotion is well suited for lattice-type and hybrid self-reconfigurable robots.

The solutions to locomotion based on self-reconfiguration are similar to the solutions outlined in the previous chapters on self-reconfiguration. However, there is an important difference. In locomotion there is no goal configuration. The shape of the robot is not important; what is important is that the robot reach the goal location in the environment through an ongoing self-reconfiguration process. Since there is no goal configuration, the robot is free to choose configurations that are suitable for ongoing locomotion.

The most common way to produce cluster-flow locomotion in self-reconfigurable robots with motion constraints is to use a uniform configuration. In this configuration, modules are arranged so that it is easy to disassemble the tail of the robot, travel the length of the robot, and finally extend its head. Often this is so simple that only local rules are needed to control the process [85, 70]. Rather than passing modules forward, it is possible to pass holes backward [84]. The effect is of course the same, but it may be easier to implement the latter and in addition it has the advan-

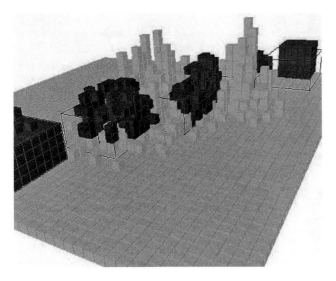

**Figure 8.1**
A cubic, simulated modular robot uses cluster-flow locomotion. The wire-frame boxes represent a moving target that the robot is trying to reach. (Courtesy of Fitch, © 2008 Sage Publications)

tage that modules only move inside the configuration and therefore the cross-section of the configuration stays the same, which is important in tight environments. On top of these low-level controllers, it is possible to add a global centralized planner that enables the robot to follow a given path in the environment [152, 153].

In systems with few motion constraints, it becomes even easier to implement cluster-flow locomotion. In these systems a uniform configuration is not needed to make control of the self-reconfiguration process easier. Instead, the configuration emerges as the robot adapts its shape to the surface on which it is traveling, allowing the robot to cross unstructured and uneven surfaces [64, 16, 17, 39] (figure 8.2).

In these implementations of cluster-flow locomotion, the configuration is used either to reduce motion constraints or to enable the robot to adapt to the environment. In essence, this means that since we are not focused on specific goal configurations, we can choose configurations that make it possible to either simplify the locomotion problem or make the robot more capable.

## 8.2 Task-Driven Growth

In chapter 7 we saw how growth processes can be used to control self-reconfiguration. In that chapter, the growth process was guided by a precise representation of the goal configuration. However, we can loosen this requirement and rather than insisting on a specific goal configuration, we are satisfied as long as the

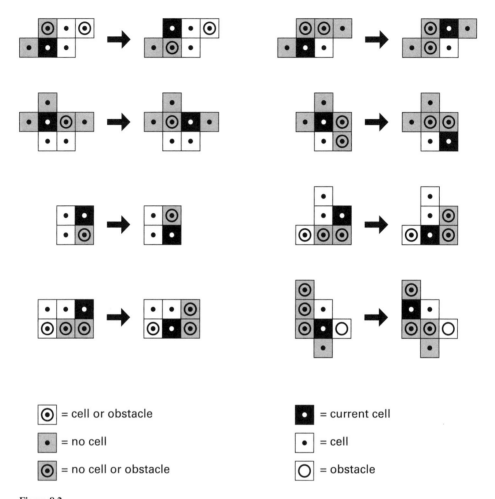

**Figure 8.2**
This is a rule set that is enough to make two-dimensional modules able to produce cluster-flow locomotion even with obstacles present. (Courtesy of Butler, © 2001 IEEE)

final configuration belongs to a class of configurations suitable for a specific task. This gives the controller more freedom to choose a configuration and thus makes the self-reconfiguration problem easier.

We may represent the class of goal configurations as a rule set [6] (figure 8.3). For instance, we can make a rule set that creates a chain of five modules and then branches in a random direction, restarts by building a new chain from this branch, and then branches again and so on. This rule set will generate a porous configuration, which for a large number of modules may be structurally very stable. That is,

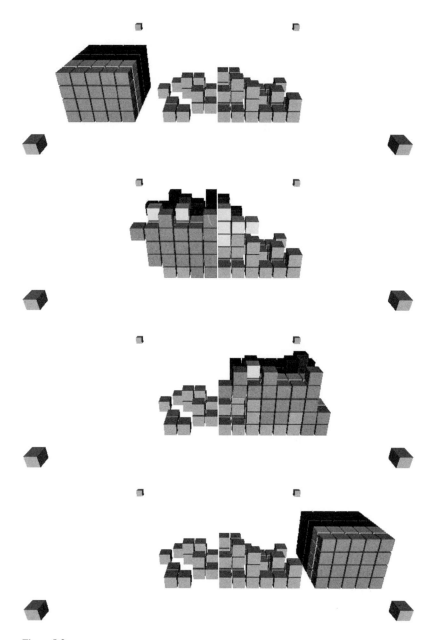

**Figure 8.3**
This is an example of a locomotion sequence controlled by the rule set shown in figure 8.2. Each layer in the z-direction acts independently. (Courtesy of Butler, © 2001 IEEE)

the class of configurations generated by this rule set may be useful for supporting a collapsing roof. Bojinov et al. [7] describe how this approach can be extended to a grasping task and to dynamically supporting a surface. This bottom-up approach is powerful, simple, and only uses local information, but it may not be trivial to design a rule set with a given complex task in mind.

Let us look at a specific example. Bojinov et al. [7] use two primitives that we presented earlier: gradients (they refer to them as scents) and growth. The basic idea behind this example is that modules are given a general idea of the direction of the object, but not anything regarding precise position, shape, or size. The configuration grows in the general direction of the object and once a module touches the object, modules start to engulf it. The rule set running on the modules contains references to six modes: SLEEP, SEARCH, TOUCH, TOUCHSEED, and FINAL.

*SLEEP* All modules start in SLEEP mode except for eight modules (fingers), which start in SEED mode.

*SEED* Modules in the mode control the growth direction and emit gradients.

*TOUCHSEED* This is similar to SEED except that modules in this mode are touching the object to be grasped.

*FINAL* Modules in SEED mode eventually change to FINAL mode when other modules closer to the object become SEEDs.

*TOUCH* This is equivalent to FINAL except that it is modules in TOUCHSEED that become TOUCH modules.

Given this intuition, the modules continually apply the following rule set (slightly modified from the original version):

· If in SLEEP mode, if a gradient is detected, go to SEARCH mode.

· If in SEARCH mode, propagate gradient, and move along the gradient unless a seed *s* has been found, in which case:

· If touching the object, then set *s* to TOUCH mode, and go to TOUCHSEED mode.

· If not touching the object, and *s* is in SEED mode, find a spot next to the seed in the direction of the object, set *s* to FINAL mode, and go to SEED mode.

· If in TOUCHSEED or SEED mode, emit scent.

· If in TOUCH or FINAL mode, propagate scent.

The effect of the rule set is that modules wander in the direction of the seeds (either TOUCHSEED or SEED). Once a seed is reached, the seed is turned off and the newly arrived module becomes the seed. The result is a process that grows eight fingers around the object in a random way, as shown in figure 8.4. Overall, this

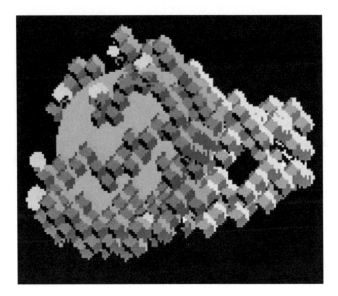

**Figure 8.4**
This shows a simulated self-reconfigurable robot engulfing an object. (Courtesy of Bojinov, © 2000 IEEE)

approach demonstrates the potential power of abandoning tightly specified goal configurations and instead focuses on the configurations with the right properties for a given task.

## 8.3 Self-Reconfiguration as a Side Effect

We have so far seen the self-reconfiguration problem from two viewpoints: the algorithmic problem of searching for a sequence of moves that can reconfigure the robot from its initial configuration to its goal configuration, and the control problem of coordinating the movements of modules in a group in such a way that the desired goal configuration is achieved. From these points of view, the self-reconfiguration problem is difficult, as we discussed at length in section 5.2. However, the question is whether there is a way to sidestep it. Can we in some way insist that the problem is not difficult and thereby make it go away? In fact, we can! At least to some degree.

The trick is simple but important because it reminds us that a solution to the self-reconfiguration problem represents a balance among task, simplifications, module design, and control. In other words, if we change the module design, our control will change as a result. The search and control views build on the assumption that self-reconfiguration can be seen as a sequence of discrete disconnections, movements, and connections. What would happen if we changed this underlying assumption? Let us assume that connections happen when modules bump into each other; disconnections happen when forces are strong enough to pull modules apart; and modules are spheres that can expand or contract. We can realize this type of connection by using unisex Velcro or weak glue. This change in basic assumptions completely changes the self-reconfiguration problem, which can now be viewed as the question of how to coordinate module expansions and contractions to make the goal configuration emerge. This is not a trivial problem either, but at least it is a completely different approach to the self-reconfiguration problem.

However, we can be more radical than this because we can insist that knowing the goal configuration a priori is an invalid assumption. The self-reconfiguration problem then becomes one of how to coordinate module oscillations to make the robot solve its task. Ishiguro et al. [50] show how this can be done in two dimensions in the task of moving a robot toward a light source (see figure 8.5). In the process, the robot encounters an obstacle, extends around it, and rejoins on the other side. In this view, the specific configurations that the robot goes through to solve this task are just a side effect of the robot pursuing its task. In fact, pursuing the task is also just a good side effect of modules making oscillations locally. While promising, this approach remains largely unexplored. In particular, it is not clear how to extend it to three dimensions in environments where gravity has to be taken into account.

## 8.4   New Challenges in Self-Reconfiguration

In this chapter and the previous two we looked at different ways to realize self-reconfiguration from a software point of view. Before we go on to the topic of controlling robots in fixed configurations, we will point out some unaddressed self-reconfiguration challenges that we need to take into account in the near future. The two challenges we discuss here are the need to maintain balance during self-reconfiguration and the need to be able to do self-reconfiguration of heterogeneous robots. In addition to these two challenges, there are also the ongoing challenges of making self-reconfiguration algorithms that can help make self-reconfigurable robots more versatile, adaptive, robust, and cheap, but we will return to these challenges in chapter 10.

### 8.4.1   Balance and Self-Reconfiguration

In environments where gravity is a factor, notably the surface of any planet, it is important that a robot maintain balance during self-reconfiguration because if it does not, it may cause damage to itself or its surroundings. Unfortunately, it is a widely held assumption that gravity can be ignored.

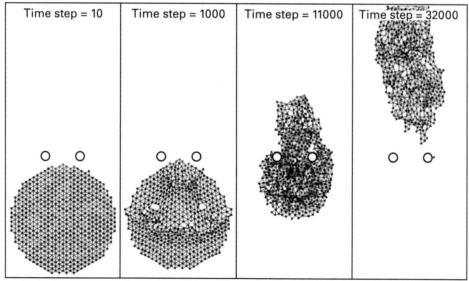

**Figure 8.5**
The Slimebot presents a completely different way to implement self-reconfigurable robots. Here modules connect softly using Velcro connectors, and self-reconfiguration is a side effect of oscillation of the modules. On the bottom is a simulation study of Slimebot approaching a light source. (Courtesy of Ishiguro)

The question is then how we perform self-reconfiguration while maintaining balance. There are two main strategies for approaching this question. If we view the self-reconfiguration problem from a search perspective, we can add a constraint that ensures that the robot maintains balance in every self-reconfiguration step. This reduces the size of the solution space: not only do we need to find a sequence of moves that reconfigure the robot from an initial to a final configuration, we also need to find a sequence that maintains the balance of the robot. This constraint is most likely to make the search problem more difficult, but it may also focus the search by eliminating a vast number of states and as a consequence make it possible to arrive at solutions faster, even though there are fewer of them.

If we view the problem of self-reconfiguration from a control point of view, the question of how to maintain balance during self-reconfiguration also may make the problem more difficult. Balance is a global property of the robot, and local controllers often have problems taking global properties into account. However, if modules are equipped with accelerometers, they can locally sense when the robot is losing its balance. This may allow them to counteract it locally.

So, in summary, it is difficult to predict if maintaining balance during self-reconfiguration is a simple or a difficult problem, but it is easy to predict that it is a problem we need to address to approach real tasks.

### 8.4.2   Heterogeneous Self-Reconfiguration

A point that has been raised by Fitch [40] is that it is unlikely that self-reconfigurable robots will remain homogeneous as we approach real tasks. As these robots mature, it is likely that some modules, while still being similar to the other modules for the purpose of self-reconfiguration, will be equipped with task-specific sensors or actuators that it is not cost-effective to put on all modules. The implication of this is that specific modules have to be in specific positions of the goal configuration. This can be viewed as yet another requirement. It is difficult to envision a strategy that can exploit this requirement in order to make the self-reconfiguration problem easier. However, application-specific sensors or actuators may potentially be used to aid the self-reconfiguration process. Advanced sensors such as cameras and laser range-finders can provide the robot with information about its surroundings and even its internal configuration and thus may be used to aid in the self-reconfiguration process.

### 8.5   Conclusion

In this chapter we have looked at self-reconfiguration solutions that are not driven by a specified goal configuration. Rather, the configuration emerges as a side effect of the robot pursuing its task.

In the context of cluster-flow locomotion, we saw that by giving up on a specific goal configuration, we can use this freedom to choose configurations that either simplify the self-configuration problem or make the robot able to adapt to the environment. In self-reconfiguration this is also the case. Bojinov et al. [7] have clearly demonstrated that control can be dramatically simplified and made adaptive if we focus on the features of the goal configuration rather than on the specific positioning of modules.

It is also possible to reduce the importance of configurations by changing the assumptions of the underlying mechanical system. If connections happen when modules bump into each other and disconnections happen when forces are strong enough to pull them apart again, self-reconfiguration is a continuous process without well-defined configurations. In these systems, configurations do not arise as a result of control but emerge as a side effect of the way modules oscillate in the system.

This is the last chapter on self-reconfiguration. In chapter 9 we will focus on controlling modular robots in fixed configurations.

## 8.6   Further Reading

Key articles are listed here under the headings of their respective sections. See the references for all articles cited in the main text.

### Locomotion through Self-Reconfiguration

E. H. Ostergaard, K. Tomita, and H. Kurokawa. Distributed metamorphosis of regular M-TRAN structures. In *Proc., 7th Int. Symp. on Distributed Autonomous Robot Systems*, pages 161–170, Toulouse, France, 2004.

E. H. Ostergaard and H. H. Lund. Distributed cluster walk for the ATRON self-reconfigurable robot. In *Proc., 8th Int. Conf. on Intelligent Autonomous Systems*, pages 291–298, Amsterdam, Netherlands, 2004.

E. Yoshida, S. Murata, A. Kamimura, K. Tomita, H. Kurokawa, and S. Kokaji. Reconfiguration planning for a self-assembling modular robot. In *Proc., IEEE Int. Symp. on Assembly and Task Planning*, pages 276–281, Fukoaka, Japan, 2001.

J. Kubica, A. Casal, and T. Hogg. Complex behaviors from local rules in modular self-reconfigurable robots. In *Proc., IEEE Int. Conf. on Robotics and Automation*, volume 1, pages 360–367, Seoul, Korea, 2001.

Z. Butler, K. Kotay, D. Rus, and K. Tomita. Cellular automata for decentralized control of self-reconfigurable robots. In *Proc., IEEE Int. Conf. on Robotics and Automation, Workshop on Modular Self-Reconfigurable Robots*, Seoul, Korea, 2001.

R. Fitch. Million module march: Scalable locomotion for large self-reconfiguring robots. *International Journal of Robotics Research*, 27(3–4):331–343, 2008.

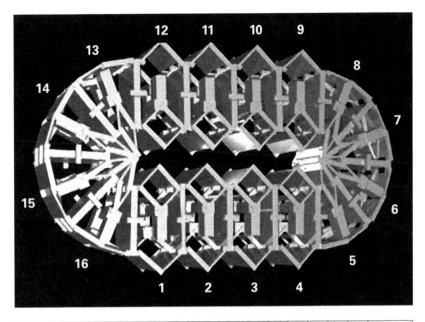

| step | 1 | 2 | 3 | 4 | 5 | 6 | 7 | 8 | 9 | 10 | 11 | 12 | 13 | 14 | 15 | 16 | trigger |
|---|---|---|---|---|---|---|---|---|---|---|---|---|---|---|---|---|---|
| 0 | → | → | → | → | ⇐ | ⇐ | ⇐ | ⇐ | → | → | → | → | ⇐ | ⇐ | ⇐ | ⇐ | 5,13 |
| 1 | ⇐ | → | → | → | → | ⇐ | ⇐ | ⇐ | ⇐ | → | → | → | → | ⇐ | ⇐ | ⇐ | 6,14 |
| 2 | ⇐ | ⇐ | → | → | → | → | ⇐ | ⇐ | ⇐ | ⇐ | → | → | → | → | ⇐ | ⇐ | 7,15 |
| 3 | ⇐ | ⇐ | ⇐ | → | → | → | → | ⇐ | ⇐ | ⇐ | ⇐ | → | → | → | → | ⇐ | 8,16 |
| 4 | ⇐ | ⇐ | ⇐ | ⇐ | → | → | → | → | ⇐ | ⇐ | ⇐ | ⇐ | → | → | → | → | 9,1 |
| 5 | → | ⇐ | ⇐ | ⇐ | ⇐ | → | → | → | → | ⇐ | ⇐ | ⇐ | ⇐ | → | → | → | 10,2 |
| 6 | → | → | ⇐ | ⇐ | ⇐ | ⇐ | → | → | → | → | ⇐ | ⇐ | ⇐ | ⇐ | → | → | 11,3 |
| 7 | → | → | → | ⇐ | ⇐ | ⇐ | ⇐ | → | → | → | → | ⇐ | ⇐ | ⇐ | ⇐ | → | 12,4 |
| 8 | → | → | → | → | ⇐ | ⇐ | ⇐ | ⇐ | → | → | → | → | ⇐ | ⇐ | ⇐ | ⇐ | 13,5 |

**Figure 9.2**
The PolyPod robot doing rolling-track locomotion using the gait control table shown (table and picture
are from Yim [139]). (Courtesy of Yim)

**Table 9.1**
Summary of rules for hormone-based control to obtain earthworm gait

| Step | Action | Resulting Hormone | Trigger Hormone | Action | Resulting Hormone |
|------|--------|-------------------|-----------------|--------|-------------------|
| 1 | ⇓ | A | D | ⇓ | A |
| 2 | ⇑ | B | A | ⇑ | B |
| 3 | ⇑ | C | B | ⇑ | C |
| 4 | ⇓ | D | C | ⇓ | D |
| (a) | | | (b) | | |

Notes: (a) This table shows the step sequence of the initiator module. The initiator performs the actions one after the other and at the beginning of each step sends the indicated hormone to the next module in the chain. (b) This table shows the set of rules that body modules use. Upon receiving a hormone, a body module performs the corresponding action and sends a new hormone to the following body module.

dependence of gait control tables on IDs [104]. The key insight is that often in gait control tables, modules go through identical sequences of steps. The only difference is that the sequence of one module is delayed a fixed number of steps compared with a neighboring module. For example, in the rolling-track gait, the motion of module $n$ at step $t$ is identical to the motion of module $n + 1$ at step $t + 1$. That is, the motion sequences are identical except that they are delayed one step. A similar observation can be made for the earthworm gait but here the following module is one step ahead. Alternatively, owing to the cyclic nature of the table, it can also be viewed as delayed three steps. This insight can be exploited to make a distributed controller.

Let us reuse the earthworm gait as an example. For this gait, the eight modules are connected in a chain. The front module is selected as the initiator of the gait. This initiator module performs a sequence of steps that corresponds to a column in the gait control table. In addition, at the beginning of each step it sends a message, a hormone, to the next module in the chain. This hormone contains information about the step that the initiator module is currently performing. Based on this information, the following module chooses the next step and passes on a hormone to the following module informing it about its choice and so on. This process continues until a hormone reaches the last module of the chain. The end result is that all modules are synchronized and are one step ahead of the neighboring module closer to the head. Table 9.1 summarizes the sequence of steps the initiator performs and the rule set used to generate hormones in this example.

The result of running the hormone-based version of the earthworm gait is almost identical to the result that would have been obtained using centralized gait control tables. However, the hormone-based implementation has two advantages: modules stay synchronized and modules can be added or removed from the chain. Modules stay synchronized because upon starting an action, the module at the head of the pened, as if it was the initiator. In other words, it never waits because this can disrupt the flow of motion. It is perfectly acceptable to miss a synchronization message because another one will be sent in the following period. Even if this one is missed too, it does not matter much because it takes a while before modules get seriously out of synchronization [120]. A side effect of this is that in role-based control, modules synchronize over time. If a chain is $n$ modules long, it takes at least $nd$ steps before the entire robot is synchronized.

The third feature is that there is no initiator—all modules are doing exactly the same thing. The head module emerges as a kind of initiator simply because there is no module in front of it to send synchronization messages. This implies that role-based control rests on the assumption that the configuration forms a tree, because loops will disrupt this automatic selection of an initiator. In loops, an initiator selection algorithm has to run before role-based control will work. An example of how this can be done using randomly assigned IDs can be found in Stoy et al. [117].

The difference between hormone-based control and role-based control is relatively small, in particular in the later implementations of hormone-based control [106]. However, modules in role-based control tend to be a little more proactive and independent compared with modules in hormone-based control, which do not move unless told to do so by a module further up the configuration. This increases the robustness and efficiency of role-based control at the cost of complete control prioritized by hormone-based control. Although we do not present them here, we should also mention that Phase Automata form another control method similar to role-based control and hormone-based control that tries to generalize these control methods [157].

### 9.1.4 Distributed Control of Complex Gaits

In the previous section we saw how a simple gait such as the earthworm gait can be implemented using gait control tables, hormone-based control, and role-based control. The question is how to implement gaits where modules perform different motions, depending on their position in the configuration. For example, in a legged robot, some modules function as spine and some modules function as legs. The answer is straightforward: a module selects its function based on the local configuration and if that is not enough, then it uses its parent's function, which is communicated through a message.

Let us look at the example of a quadrupedal walker. The idea behind the implementation is the same in role-based and hormone-based control [118, 106] and in both cases relies on the configuration being a tree: the delay between each pair of legs should be half a period. When the front leg moves forward, the rear leg should move backward and the other way around. The same is true for the relationship

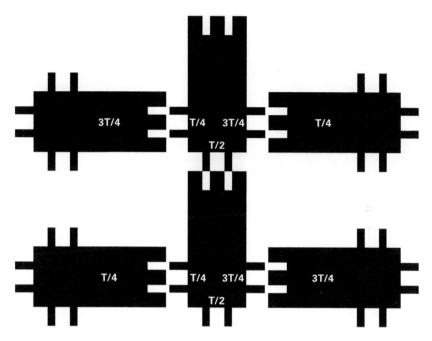

**Figure 9.4**
This figure shows six CONRO modules, which are represented by black boxes, connected to form a quad-rupedal walker. The delay positioned at each of the male connectors is the delay of the motion of the module attached to that connector. The result is that the motion of the two front legs, the two rear legs, and each pair of legs is half a period apart. (Courtesy of Stoy, © 2002 IEEE)

between the left and the right leg within a pair. In order to achieve this, the delays of role-based control can be arranged as shown in figure 9.4.

With the delays fixed, we need to describe the motions. Intuitively, the legs should be lifted from the ground when moving forward and touching the ground when moving backward. We use the following motion equation for the left legs:

$$A(\text{leftleg}, t) = \begin{cases} \text{pitch}(t) = 35° \cos\left(\frac{2\pi}{T} t\right) - 55° \\ \text{yaw}(t) = 40° \sin\left(\frac{2\pi}{T} t\right). \end{cases} \tag{9.2}$$

The equation for the right legs is obtained by replacing $t$ by $2\pi - t$, giving the same motion but in the opposite direction. The spine module between two pairs of legs should bend from side to side to increase the length of each step. The parameters for this motion are

$$A(\text{spine}, t) = \begin{cases} \text{pitch}(t) = 0° \\ \text{yaw}(t) = 25° \cos\left(\frac{2\pi}{T} t + \pi\right). \end{cases} \tag{9.3}$$

**Figure 9.5**
The CONRO robot in a hexapod configuration. (Courtesy of Stoy, © 2002 IEEE)

For simplicity we pick the same period $T$ for all roles. The parameter $T$ can later be used to control the locomotion speed. We have now specified the delays and the motions; all that is left is to provide modules a way to decide which function they should perform. In this example we do it based on the local configuration. In particular, it is done based on which connector of the parent the module is connected to. This information is communicated to the module through a message from the parent communicated at the same time as the synchronization signal.

This implementation of quadrupedal walking on the CONRO robot achieves a speed of 13.8 cm/second. The robot can also be extended with an extra pair of legs, as shown in figure 9.5, without requiring any changes to the code or decrease in performance. With the addition of a sidewinder role and a slightly more complex role selection function, it is also possible for the robot to change gait, depending on the configuration, as shown in figure 9.6 (see Stoy et al. [121] for details).

### 9.1.5 Using Sensors in Closed-Loop Control

We have described a number of methods that can be used to implement complex gaits in self-reconfigurable robots. However, we have so far only described how to generate an open-loop gait; that is, gaits that do not take the environment into account.

One approach is to add another layer that, based on knowledge about the environment, can plan a path through the environment. Eldershaw and Yim [38] and Yoshida et al. [151] pursued this approach using a probabilistic road-map planner and an

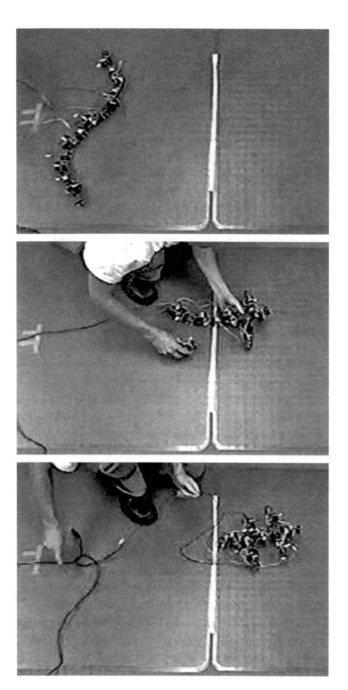

**Figure 9.6**
The CONRO robot first covers a distance of 63 cm using a sidewinder gait (top). The robot is then manually reconfigured into a quadrupedal walker (middle). Finally, the robot walks 87 cm (bottom). Note that the cables only provide power. (Courtesy of Stoy, © 2003 IEEE)

M. Yim, D. Duff, and Y. Zhang. Closed-chain motion with large mechanical advantage. In *Proc., IEEE/RSJ Int. Conf. on Intelligent Robots and Systems*, pages 318–323, Maui, Hawaii, 2001.

D. J. Christensen, D. Brandt, and K. Stoy. Towards artificial ATRON animals: Scalable anatomy for self-reconfigurable robots. In *Proc., RSS Workshop on Self-Reconfigurable Modular Robots*, pages 1–2, Philadelphia, PA, 2006.

J. Campbell and P. Pillai. Collective actuation. *International Journal of Robotics Research*, 27(3–4):299–314, 2008.

M. Yim, J. Reich, and A. Berlin. Two approaches to distributed manipulation. In H. Choset and K. Bohringer, editors, *Distributed Manipulation*. Kluwer Academic Publishing, Norwell, MA, 2000.

J. Kubica, A. Casal, and T. Hogg. Agent-based control for object manipulation with modular self-reconfigurable robots. In *Proc., Int. Joint Conf. on Artificial Intelligence*, pages 1344–1352, Seattle, WA, 2001.

# 10 Research Challenges

Self-reconfigurable robots have come a long way since the conception of the idea by Fukuda in the mid-1980s. The field has in particular picked up momentum in the past decade with an increase in the number of active researchers and maybe as a consequence significant research results. For example, self-reconfigurable robot-related work won best paper award at the IEEE International Conference on Robotics and Automation in 2006 [92] and the 8th International Symposium on Distributed Autonomous Robotic Systems in 2006 [68]. Nevertheless, there are still challenges that the research community has to address. Judging from the wealth of examples of different locomotion patterns implemented on pack robots, it appears that pack robots are maturing and moving closer and closer to application. At the other extreme, swarm robots, while being significantly less mature, are breaking new ground by raising fundamentally new basic research questions, such as how to program and control massive swarm robots and how to design and physically build swarms of modules. In this chapter we will mainly look at the challenges involved in moving self-reconfigurable robots toward application, with a focus on pack and to some degree herd robots. The reason is that the fundamental questions we need to answer to make swarm robots useful are less clear and that it is not even clear if the goal should be to make them useful. The real goal of swarm robots may instead be to make contributions to basic research that can be exploited in completely different fields, such as software engineering.

This chapter will proceed as follows. We will discuss the complexity of real tasks and how we can use the special abilities of self-reconfigurable robots to deal with this complexity. Furthermore, we will introduce concepts from behavior-based robotics [2, 71] that allow us to further reduce the complexity of real tasks by looking at subtasks in isolation. We will also briefly discuss some of the changes needed with regard to hardware and finally end by integrating all the functionalities described here into a conceptual framework that may be useful when we approach real tasks.

## 10.1   Facing the Complexity of Real Tasks

The first challenge that will face us when we move toward real tasks is the fundamental problem that real-world applications are difficult for robots in general. There are many reason for this. One part of the problem is the limitations of sensors, actuators, and even materials. Another part is that even with sufficient hardware, an understanding of how to develop intelligent controllers is lacking. We are able to develop robots for highly structured, known environments, such as factory assembly lines, but developing a robot to go grocery shopping is beyond our capabilities.

Self-reconfigurable robots are probably unlikely to change this picture dramatically. Nevertheless, they do represent a new approach because, as opposed to conventional robots, they have a flexible bodily structure. So not only can a self-reconfigurable robot optimize its controller for a specific task, it can also optimize its bodily structure. This is important because the behavior of a robot, and thus its ability to perform a task, is the result of an interaction among environment, body, and brain [90]. Let us look at an example to make this clear. Suppose the task is to walk down a staircase. An example of a bodily structure optimized for this task is the Slinky spring. In fact, it is optimized to a degree where its body alone can solve the task and thus there is no need for a controller. The bodily structure of the Slinky spring, however, is unfit for most other tasks, so if we want a more versatile bodily structure we have to find an alternative. An example of a more versatile bodily structure is that of a humanoid robot, but since its body is less optimized for the task, it has to rely on advanced sensors, powerful and precise actuators, and a complex controller to walk down a staircase. If we compare the two solutions, the Slinky solution is much simpler, but even though this solution is simpler, the humanoid robot may be a better choice overall because of its versatility. That is, using conventional robots we are often forced to make a trade-off between simplicity and versatility.

A self-reconfigurable robot can optimize its bodily structure to make every task it encounters as easy as possible to perform. This may allow self-reconfigurable robots to perform a wide range of tasks and at the same time make the solutions to individual tasks simple. We therefore avoid the trade-off of conventional robots and may obtain solutions that are both simple and versatile. Moreover, since an optimized bodily structure is the key to obtaining simple solutions, it may even be possible for self-reconfigurable robots to do tasks that are too complex for conventional robots because the bodily structures of conventional robots are less optimized for tasks that do not lend themselves to solutions based on a robot arm. So while real tasks certainly will challenge our hardware and software solutions, they may also present a way to demonstrate the power of self-reconfigurable robots.

**Figure 10.1**
A typical organization of a behavior-based controller for navigation.

## 10.2   From Basic Functionalities to Behaviors

In order to perform a real task, a self-reconfigurable robot needs access to a range of basic functionalities. It may need to move around in the environment; it may need to manipulate objects; and it may also need to change shape. This opens two challenges for researchers in the field of self-reconfigurable robots. First, we need to improve the basic functionalities of these robots to make them suitable for the complexity of real tasks. Second, we need a way to combine these basic functionalities into a coherent controller. In this section we look at the basic functionalities and in the following sections the questions of how to adapt them and finally how to combine them.

Self-reconfigurable robots are able to perform stereotypical tasks. These tasks are characterized by the fact that they capture only some of the aspects of the real task from which they were derived. One of the aspects that is often forgotten is the complexity of the environment and the need to respond to it based on sensed information. In order to perform a real task, a self-reconfigurable robot needs to be equipped with more than an open-looped controller. It needs a closed-looped controller that allows it to interact with its environment and the objects of interest for the task. It needs something akin to the behaviors of behavior-based robotics [2, 71], but with more focus on the distributed nature of self-reconfigurable robots. A behavior-based controller is composed of a number of behaviors, with each focusing on solving one aspect of a task. For example, in navigation, different behaviors may be responsible for moving the robot, avoiding obstacles, navigating between key points in the environment, and planning a path through key points that takes the robot to its destination, as illustrated in figure 10.1. An important thing to note is that behaviors do not enforce a specific way of implementing them and we may therefore use any of the methods presented in this book to provide the implementation of a behavior. The important thing about behaviors is that they give us an alternative way to simplify real tasks to make them manageable. Rather than over-simplifying the task itself, behaviors simplify the task by splitting it up into small, well-defined subtasks. We think this is an important idea that is needed to take self-reconfigurable robots beyond stereotypical tasks. This will allow us to look at different aspects of a real task in isolation. For instance, we may implement a behavior that is responsible for maintaining the balance of a robot during self-reconfiguration. On top of this behavior we may implement one that adapts the robot's bodily structure to its task without considering the question of balance because this task is handled by the lower-level behavior.

## 10.3   Behavior Adaptation

As we already mentioned, one of the powerful features of self-reconfigurable robots is that they can adapt both their bodily structure and their controller to a task. So far we have discussed the potential of using this ability to simplify otherwise complex, real tasks and use behaviors to simplify them even more by looking at subtasks in isolation. In addition, this ability can also be used to optimize a behavior for a specific task. For example, if we look at the task of moving a robot forward across difficult terrain, we may choose to implement a solution in the form of a walking gait. During execution, the terrain becomes less difficult and thus it would be beneficial if the robot could increase its speed in response. A self-reconfigurable robot can do this in two ways. It can increase its speed by increasing the frequency of the gait or it may also extend its legs and thus make each step longer and in this way increase its speed.

There are a few studies of behavior adaptation of self-reconfigurable robots in the literature, but they tend to adapt either the controller or the body, but not both at the same time and only in the context of stereotypical tasks. One of the most investigated is cluster-flow locomotion, in which the robot moves forward while continually adapting its body to the environment [19]. There are also examples of controller adaptation. For example, Kamimura et al. [55] demonstrate how a walking gait can be adapted and optimized to allow a walker to climb an incline.

Behavior adaptation is characterized by not changing the behavior radically. It allows the robot to adapt and optimize its behavior for aspects of a task that were not known at the time of implementation. For example, a robot can adapt its behavior to handle an increase in payload, a change in the slope of an incline, or maybe a few module failures. We think that once we have implemented behaviors for real subtasks, it becomes important to study how we can adapt these behaviors online. The degree to which we potentially can adapt behaviors is unique to self-reconfigurable robots and suggests a way to realize the features of robustness and adaptability and thus gives us yet another way to demonstrate the power of these robots.

## 10.4  Behavior Selection

Behavior adaptation potentially provides self-reconfigurable robots with adaptability and robustness, but this adaptation alone does not provide the last feature of self-reconfigurable robots: versatility. Furthermore, behavior adaptation can only accommodate a certain degree of change in the environment. If the changes are dramatic enough, behavior adaptation is not enough to ensure that a robot can continue to perform its task. For example, if the tunnel through which the robot is moving suddenly narrows, we cannot simply adapt the walking gait; the robot needs to completely change its behavior, for instance, by changing to a snakelike locomotion pattern. The consequence is that in addition to behavior adaptation, we need some way to completely change the behavior of the robot. One way is through behavior coordination. Where adaptation is internal to individual behaviors and thus limited, behavior coordination deals, as the word suggests, with coordination of behaviors in terms of which are active and how they interact. At this level of control we may provide self-reconfigurable robots with the versatility needed to handle a wide range of tasks and dramatic changes in the environment. Behavior coordination allows us to collect individual behaviors into a coherent control system that allows a robot to perform its task under changing circumstances.

In behavior-based controllers, all behaviors run in parallel, but most often only one behavior controls the robot at any given time. The behavior that controls the robot is the behavior with the highest priority unless this behavior doesn't need to do anything in relation to its task, in which case control is passed on to the behavior with the second-highest priority and so on. For example, the high-priority obstacle avoidance behavior controls the robot when obstacles are nearby, but otherwise lets lower-priority behaviors such as navigation control the robot. This process of choosing which behavior is active is called behavior selection. Behavior selection is not limited to selecting only one behavior at a time; sometimes more behaviors can control the robot at the same time if their commands do not conflict or can be combined in a meaningful way. Behavior selection is a continuous process that monitors sensors or even the internals of some behaviors to decide which behaviors should be active. In this sense behavior selection is similar to behavior adaptation. However, behavior selection can also decide to replace some behaviors with others while operating and as a result completely change the behavior mode of the robot.

Behavior selection is often based on sensed information about the environment or changes in a task. In the context of self-reconfigurable robots, some preliminary work has demonstrated how this can be done. Stoy et al. [119] use sensor input to select which roles specific modules in the configuration should play. For example, upon detecting an obstacle, the controller activates some roles as a result of which

It also becomes increasingly clear that as we move closer to real applications, the age of homogeneous self-reconfigurable robots may come to an end. Often specific functionality is needed for specific applications that cannot be achieved using a general-purpose module. This may violate the vision of a self-reconfigurable robot, which is to have homogeneous modules from which all other robots, in theory, may be built. However, this may not necessarily be the case as long as the level of reuse of most of the modules is high. For example, even though our extraterrestrial, self-reconfigurable, geology robot has several different modules optimized for drilling, digging, etc., this is acceptable as long as the bulk of the robot is made from general-purpose modules that can be reused in other application contexts.

## 10.8    Conclusion

In this final chapter, we have argued that we have the tools to approach real tasks rather than stereotypical, isolated tasks. We have argued that we should exploit the ability of self-reconfigurable robots to change their bodily structure to make the complexity of real tasks more manageable. We have also suggested that a behavior-based robotics-inspired approach can reduce the complexity further by allowing us to look at subtasks in isolation and then later combine the resulting behaviors into coherent global behavior modes through behavior selection. Finally, we have described how global behavior modes provide a convenient framework for understanding how we can tie all the isolated functionalities developed in the field over the past twenty years into coherent controllers capable of handling real tasks and thereby realize the vision of self-reconfigurable robots.

We hope that after reading this book the reader agrees that the field of self-reconfigurable robots has made significant progress. We have gained a good understanding of the different trade-offs among different mechanical solutions. We have also developed an insight into the interaction between hardware and control, specifically in the context of self-reconfiguration.

We find that this is an exciting time for self-reconfigurable robots because we have solved some of the fundamental problems and are ready and confident enough to on one hand, move pack robots toward application and on the other, to address new basic research questions regarding realization and control of herd and swarm robots.

## 10.9    Further Reading

Key articles are listed under the headings of their respective sections. See the references for all articles cited in the main text.

## Research Challenges

M. Yim, W.-M. Shen, B. Salemi, D. Rus, M. Moll, H. Lipson, E. Klavins, and G. S. Chirikjian. Modular self-reconfigurable robot systems. In *IEEE Robotics & Automation*, 14(1):43–52, 2007.

R. Pfeifer and C. Scheier. *Understanding Intelligence*. MIT Press, Cambridge, MA, 1999.

R. C. Arkin. *Behavior-Based Robotics*. MIT Press, Cambridge, MA, 1998.

Maja J. Matarić. Behavior-based control: Examples from navigation, learning, and group behavior. *Journal of Experimental and Theoretical Artificial Intelligence*, 9(2–3):323–336, 1997.

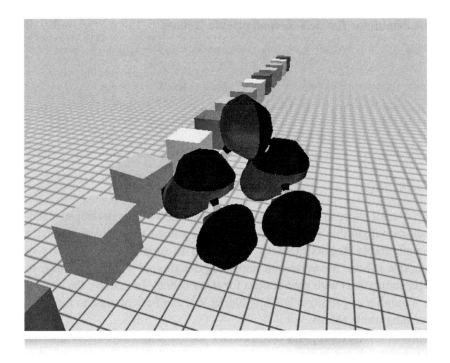

**Figure A.1**
Simulation of ATRON and Odin in USSR.

# References

[1] S. K. Agrawal, L. Kissner, and M. Yim. Joint solutions of many degrees-of-freedom systems using dextrous workspaces. In *Proc., IEEE Int. Conf. on Robotics and Automation*, volume 3, pages 2480–2485, Seoul, Korea, 2001.

[2] R. C. Arkin. *Behavior-Based Robotics*. MIT Press, Cambridge, MA, 1998.

[3] R. Beckers, O. E. Holland, and J. L. Deneubourg. From action to global task: Stigmergy and collective robotics. In *Proc., Artificial Life 4*, pages 181–189, Cambridge, MA, 1994.

[4] G. Beni and J. Wang. Theoretical problems for the realization of distributed robotic systems. In *Proc., IEEE Conf. on Robotics and Automation*, pages 1914–1920, 1991.

[5] J. Bishop, S. Burden, E. Klavins, R. Kreisberg, W. Malone, N. Napp, and T. Nguyen. Self-organizing programmable parts. In *Proc., Int. Conf. on Intelligent Robots and Systems*, pages 3684–3691, Edmonton, Alberta, Canada, 2005.

[6] H. Bojinov, A. Casal, and T. Hogg. Emergent structures in modular self-reconfigurable robots. In *Proc., IEEE Int. Conf. on Robotics and Automation*, volume 2, pages 1734–1741, San Francisco, CA, 2000.

[7] H. Bojinov, A. Casal, and T. Hogg. Multiagent control of self-reconfigurable robots. In *Proc., Fourth Int. Conf. on Multiagent Systems*, pages 143–150, Boston, MA, 2000.

[8] E. Bonabeau, M. Dorigo, and G. Theraulaz. *Swarm Intelligence: From Natural to Artificial Systems*. Santa Fe Institute Studies in the Sciences of Complexity Proceedings. Oxford University Press, 1999.

[9] D. Brandt. Comparison of A* and RRT-connect motion planning techniques for self-reconfiguration planning. In *Proc., IEEE/RSJ Int. Conf. on Intelligent Robots and Systems*, pages 892–897, Beijing, China, 2006.

[10] D. Brandt and D. J. Christensen. A new meta-module for controlling large sheets of ATRON modules. In *Proc., IEEE/RSJ Int. Conf. on Intelligent Robots and Systems*, pages 2375–2380, San Diego, CA, 2007.

[11] D. Brandt and E. H. Ostergaard. Behaviour subdivision and generalization of rules in rule-based control of the ATRON self-reconfigurable robot. In *Proc., Int. Sym. on Robotics and Automation*, pages 67–74, Querétaro, Mexico, 2004.

[12] Z. Butler, S. Byrnes, and D. Rus. Distributed motion planning for modular robots with unit-compressible modules. In *Proc., IEEE/RSJ Int. Conf. on Intelligent Robots and Systems*, volume 2, pages 790–796, Maui, Hawaii, 2001.

[13] [13] Z. Butler, R. Fitch, and D. Rus. Distributed control for unit-compressible robots: Goal-recognition, locomotion, and splitting. *IEEE/ASME Transactions on Mechatronics, Special Issues on Self-Reconfigurable Robots*, 7(4):418–403, 2002.

[14] Z. Butler, R. Fitch, and D. Rus. Experiments in locomotion with a unit-compressible modular robot. In *Proc., IEEE/RSJ Int. Conf. on Intelligent Robots and Systems*, pages 2813–2818, Lausanne, Switzerland, 2002.

[15] Z. Butler, R. Fitch, and D. Rus. Experiments in distributed control of modular robots. In B. Siciliano and P. Dario, editors, *Proc., Experimental Robotics VIII*, volume 5 of *Springer Tracts in Advanced Robotics*, pages 307–316. Springer, 2003.

[16] Z. Butler, K. Kotay, D. Rus, and K. Tomita. Cellular automata for decentralized control of self-reconfigurable robots. In *Proc., IEEE Int. Conf. on Robotics and Automation, Workshop on Modular Self-Reconfigurable Robots*, Seoul, Korea, 2001.

[17] Z. Butler, S. Murata, and D. Rus. Distributed replication algorithms for self-configuring modular robots. In *Proc., Distributed Autonomous Robotic Systems 5*, pages 37–48, Fukuoka, Japan, 2002.

[18] Z. Butler and D. Rus. Distributed motion planning for 3-D modular robots with unit-compressible modules. In *Proc., Workshop on the Algorithmic Foundations of Robotics*, pages 435–451, Nice, France, 2002.

[19] Z. Butler and D. Rus. Distributed locomotion algorithms for self-configurable robots operating on rough terrain. In *Proc., IEEE Int. Symp. on Computational Intelligence in Robotics and Automation*, pages 880–885, Kobe, Japan, 2003.

[20] Z. Butler and D. Rus. Distributed planning and control for modular robots with unit-compressible modules. *International Journal of Robotics Research*, 22(9):699–715, 2003.

[21] J. Campbell and P. Pillai. Collective actuation. *International Journal of Robotics Research*, 27(3–4):299–314, 2008.

[22] J. Campbell, P. Pillai, and S. C. Goldstein. The robot is the tether: Active, adaptive power routing for modular robots with unary inter-robot connectors. In *Proc., IEEE/RSJ Int. Conf. on Intelligent Robots and Systems*, pages 4108–4115, Edmonton, Alberta, Canada, 2005.

[23] A. Casal and M. Yim. Self-reconfiguration planning for a class of modular robots. In G. T. McKee and P. S. Schenker, editors, *Proc., Sensor Fusion and Decentralized Control in Robotic Systems II*, volume 3839, pages 246–257. SPIE, Bellingham, WA, 1999.

[24] A. Castano, W.-M. Shen, and P. Will. CONRO: Towards deployable robots with inter-robot metamorphic capabilities. *Autonomous Robots*, 8(3):309–324, 2000.

[25] C.-J. Chiang and G. S. Chirikjian. Modular robot motion planning using similarity metrics. *Autonomous Robots*, 10(1):91–106, 2001.

[26] G. Chirikjian, A. Pamecha, and I. Ebert-Uphoff. Evaluating efficiency of self-reconfiguration in a class of modular robots. *Robotics Systems*, 13:317–338, 1996.

[27] G. S. Chirikjian. Kinematics of a metamorphic robotic system. In *Proc., IEEE Int. Conf. on Robotics and Automation*, volume 1, pages 449–455, San Diego, CA, 1994.

[28] G. S. Chirikjian and J. W. Burdick. Geometric approach to hyper-redundant manipulator obstacle avoidance. *Journal of Mechanical Design—Transactions of the ASME*, 114(4):580–585,1992.

[29] G. S. Chirikjian, Y. Zhou, and J. Suthakorn. Self-replicating robots for lunar development. *IEEE/ASME Transactions on Mechatronics*, 7(4):462–472, 2002.

[30] L. Chittka. Dances as windows into insect perception. *PLoS Biology*, 2(7):e216, 2004.

[31] H. C. H. Chiu, M. Rubenstein, and W.-M. Shen. Multifunctional SuperBot with rolling track configuration. In *Proc., IROS 2007 Workshop on Self-Reconfigurable Robots, Systems and Applications*, San Diego, CA, 2007.

[32] D. J. Christensen. Evolution of shape-changing and self-repairing control for the ATRON self-reconfigurable robot. In *Proc., IEEE Int. Conf. on Robotics and Automation (ICRA)*, pages 2539–2545, Orlando, FL, 2006.

[33] D. J. Christensen. Experiments on fault-tolerant self-reconfiguration and emergent self-repair. In *Proc., Symp. on Artificial Life part of the IEEE Symposium Series on Computational Intelligence*, pages 355–361, Honolulu, Hawaii, 2007.

[34] D. J. Christensen, D. Brandt, and K. Stoy. Towards artificial ATRON animals: Scalable anatomy for self-reconfigurable robots. In *Proc., RSS Workshop on Self-Reconfigurable Modular Robots*, pages 1–2, Philadelphia, PA, 2006.

[35] D. J. Christensen, E. H. Ostergaard, and H. H. Lund. Meta-module control for the ATRON self-reconfigurable robotic system. In *Proc., 8th Conf. on Intelligent Autonomous Systems*, pages 685–692, Amsterdam, Netherlands, 2004.

[36] D. J. Christensen and K. Stoy. Selecting a meta-module to shape-change the ATRON self-reconfigurable robot. In *Proc., IEEE Int. Conf. on Robotics and Automation*, pages 2532–2538, Orlando, FL, 2006.

[37] J. J. Craig. *Introduction to Robotics: Mechanics and Control*. 3rd ed. Prentice Hall, Engleword Cliffs, NJ, 2003.

[38] C. Eldershaw and M. Yim. Motion planning of legged vehicles in an unstructured environment. In *Proc., IEEE Int. Conf. on Robotics and Automation*, volume 4, pages 3383–3389, Seoul, Korea, 2001.

[39] R. Fitch. Million module march: Scalable locomotion for large self-reconfiguring robots. *International Journal of Robotics Research*, 27(3–4):331–343, 2008.

[40] R. C. Fitch. Heterogenous self-reconfiguring robotics. PhD thesis, Dartmouth College, Hanover, NH, 2004.

[41] M. Fromherz, M. Hoeberechts, and W. Jackson. Towards constraint-based actuation allocation for hyper-redundant manipulators. In *Workshop on Constraints in Control*, Alexandria, VA, 1999.

[42] M. Fromherz, T. Hogg, Y. Shang, and W. Jackson. Modular robot control and continuous constraint satisfaction. In *Proc., IJCAI Workshop on Modelling and Solving Problems with Constraints*, pages 47–56, Seattle, WA, 2001.

[43] M. Fromherz and W. Jackson. Predictable motion of hyper-redundant manipulators using constrained optimization control. In *Proc., Int. Conf. on Artificial Intelligence*, pages 1141–1147, Las Vegas, NV, 2000.

[44] T. Fukuda and S. Nakagawa. Approach to the dynamically reconfigurable robotic system. *Intelligent and Robotic Systems*, 1(1):55–72, 1988.

[45] T. Fukuda and S. Nakagawa. Dynamically reconfigurable robotic systems. In *Proc., IEEE Int. Conf. on Robotics and Automation*, volume 3, pages 1581–1586, Philadelphia, PA, 1988.

[46] R. F. M. Garcia, D. J. Christensen, K. Stoy, and A. Lyder. Hybrid approach: A self-reconfigurable communication network for modular robots. In *Proc., First Int. Conf. on Robots and Communication*, Athens, Greece, 2007.

[47] S. Hackwood and J. Wang. The engineering of cellular robotic systems. In *Proc., IEEE Int. Symp. on Intelligent Control*, pages 70–75, Arlington, VA, 1988.

[48] K. Hosokawa, T. Tsujimori, T. Fujii, H. Kaetsu, H. Asama, Y. Kuroda, and I. Endo. Self-organizing collective robots with morphogenesis in a vertical plane. In *Proc., IEEE Int. Conf. on Robotics and Automation*, pages 2858–2863, Leuven, Belgium, 1998.

[49] N. Inou, H. Kobayashi, and M. Koseki. Development of pneumatic cellular robots forming a mechanical structure. In *Proc., Seventh Int. Conf. on Control, Automation, Robotics and Vision*, pages 63–68, Singapore, 2002.

[50] A. Ishiguro, M. Shimizu, and T. Kawakatsu. Don't try to control everything!: An emergent morphology control of a modular robot. In *Proc., IEEE/RSJ Int. Conf. on Intelligent Robots and Systems*, pages 981–985, Sendai, Japan, 2004.

[51] C. Jones and M. J. Matarić. From local to global behavior in intelligent self-assembly. In *Proc., IEEE Int. Conf. on Robotics and Automation*, pages 721–726, Taipei, Taiwan, 2003.

[52] M. W. Jorgensen, E. H. Ostergaard, and H. H. Lund. Modular ATRON: Modules for a self-reconfigurable robot. In *Proc., IEEE/RSJ Int. Conf. on Robots and Systems*, pages 2068–2073, Sendai, Japan, 2004.

[53] R. A. Freitas Jr. and R. C. Merkle. *Kinematic Self-Replicating Machines*. Landes Bioscience, Georgetown, TX, 2004.

[54] A. Kamimura, H. Kurokawa, E. Yoshida, S. Murata, K. Tomita, and S. Kokaji. Automatic locomotion design and experiments for a modular robotic system. *IEEE/ASME Transactions on Mechatronics*, 10(3):314–325, 2005.

[55] A. Kamimura, H. Kurokawa, E. Yoshida, K. Tomita, S. Kokaji, and S. Murata. Distributed adaptive locomotion by a modular robotic system, M-TRAN II. In *Proc., IEEE/RSJ Int. Conf. on Intelligent Robots and Systems*, volume 2, pages 2370–2377, Sendai, Japan, 2004.

[56] A. Kamimura, H. Kurokawa, E. Yoshida, K. Tomita, S. Murata, and S. Kokaji. Automatic locomotion pattern generation for modular robots. In *Proc., IEEE Int. Conf. on Robotics and Automation*, pages 714–720, Taipei, Taiwan, 2003.

[139] M. Yim. Locomotion with a unit-modular reconfigurable robot. PhD thesis, Department of Mechanical Engineering, Stanford University, Stanford, CA, 1994.

[140] M. Yim. New locomotion gaits. In *Proc., Int. Conf. on Robotics and Automation*, pages 2508–2514, San Diego, CA, 1994.

[141] M. Yim, D. Duff, and Y. Zhang. Closed-chain motion with large mechanical advantage. In *Proc., IEEE/RSJ Int. Conf. on Intelligent Robots and Systems*, pages 318–323, Maui, Hawaii, 2001.

[142] M. Yim, D. G. Duff, and K. D. Roufas. PolyBot: A modular reconfigurable robot. In *Proc., IEEE Int. Conf. on Robotics and Automation*, pages 514–520, San Francisco, CA, 2000.

[143] M. Yim, D. Goldberg, and A. Casal. Connectivity planning for closed-chain reconfiguration. In G. T. McKee, T. Gerard, and P. S. Schenker, editors, *Proc., Sensor Fusion and Decentralized Control in Robotic Systems III*, volume 4196, pages 402–412. SPIE, Bellingham, WA, 2000.

[144] M. Yim, S. Homans, and K. Roufas. Climbing with snake-like robots. In *Proc., IFAC Workshop on Mobile Robot Technology*, Jejudo, Korea, 2001.

[145] M. Yim, J. Lamping, E. Mao, and J. G. Chase. Rhombic dodecahedron shape for self-assembling robots. Technical report, Xerox PARC, 1997. SPL TechReport P9710777.

[146] M. Yim, J. Reich, and A. Berlin. Two approaches to distributed manipulation. In H. Choset and K. Bohringer, editors, *Distributed Manipulation*. Kluwer Academic Publishing, Norwell, MA, 2000.

[147] M. Yim, B. Shirmohammadi, J. Sastra, M. Park, M. Dugan, and C. J. Taylor. Towards robotic self-reassembly after explosion. In *Proc., IEEE/RSJ Int. Conf. on Intelligent Robots and Systems*, pages 2767–2772, San Diego, CA, 2007.

[148] M. Yim, Y. Zhang, J. Lamping, and E. Mao. Distributed control for 3D metamorphosis. *Autonomous Robots*, 10(1):41–56, 2001.

[149] M. Yim, Y. Zhang, K. Roufas, D. Duff, and C. Eldershaw. Connecting and disconnecting for chain self-reconfiguration with PolyBot. *IEEE/ASME Transactions on Mechatronics*, 7(4):442–451, 2002.

[150] E. Yoshida, S. Kokaji, S. Murata, H. Kurokawa, and K. Tomita. Miniaturized self-reconfigurable system using shape memory alloy. In *Proc., IEEE/RSJ Int. Conf. on Intelligent Robots and Systems*, volume 3, pages 1579–1585, Busan, Korea, 1999.

[151] E. Yoshida, H. Kurokawa, A. Kamimura, K. Tomita, S. Kakaji, and S. Murata. Planning behaviors of a modular robot: An approach applying a randomized planner to coherent structure. In *Proc., IEEE/RSJ Int. Conf. on Intelligent Robots and Systems*, volume 2, pages 2056–2061, Sendai, Japan, 2004.

[152] E. Yoshida, S. Murata, A. Kamimura, K. Tomita, H. Kurokawa, and S. Kokaji. A motion planning method for a self-reconfigurable modular robot. In *Proc., IEEE/RSJ Int. Conf. on Intelligent Robots and Systems*, pages 590–597, Maui, Hawaii, 2001.

[153] E. Yoshida, S. Murata, A. Kamimura, K. Tomita, H. Kurokawa, and S. Kokaji. Reconfiguration planning for a self-assembling modular robot. In *Proc., IEEE Int. Symposium on Assembly and Task Planning*, pages 276–281, Fukuoka, Japan, 2001.

[154] E. Yoshida, S. Murata, A. Kamimura, K. Tomita, H. Kurokawa, and S. Kokaji. Evolutionary motion synthesis for a modular robot using genetic algorithm. *Journal of Robotics and Mechatronics*, 15(2):227–237, 2003.

[155] E. Yoshida, S. Murata, A. Kamimura, K. Tomita, H. Kurokawa, and S. Kokaji. Evolutionary synthesis of dynamic motion and reconfiguration process for a modular robot M-TRAN. In *Proc., IEEE Int. Symp. on Computational Intelligence in Robotics and Automation*, pages 1004–1010, Kobe, Japan, 2003.

[156] Y. Zhang, K. Roufas, C. Eldershaw, M. Yim, and D. Duff. Sensor computation in modular self-reconfigurable robots. In B. Siciliano and P. Dario, editors, *Proc., Experimental Robotics VIII*, volume 5 of *Springer Tracts in Advanced Robotics*, pages 276–286. Springer, 2003.

[157] Y. Zhang, M. Yim, C. Eldershaw, D. Duff, and K. Roufas. Phase automata: A programming model of locomotion gaits for scalable chain-type modular robots. In *Proc., IEEE/RSJ Conf. on Intelligent Robots and Systems*, volume 3, pages 2442–2447, Las Vegas, NV, 2003.

[158] V. Zykov, E. Mytilinaios, B. Adams, and H. Lipson. Self-reproducing machines. *Nature*, 435:163–164, 2005.

[159] V. Zykov, S. Mytilinaios, M. Desnoyer, and H. Lipson. Evolved and designed self-reproducing modular robotics. *IEEE Transactions on Robotics*, 23(2):308–319, 2007.

# Index

3D-Unit, 15, 18, 78, 79, 84

A*, 121
Active robustness, 32
Actuators, 55–64
 strength, 62
 types, 63
Adaptability, 32
Alignment of connectors, 47, 49, 65–66
Applications, potential, 22
Asimov, Isaac, 11
ATRON, 3–6, 18, 78, 79, 84
 chain-based motion of, 60
 complexity of, 37
 meta-module, 106
 module autonomy, 56
 power-sharing, 89
 size of configuration space, 104
Attractor, 130–133
Autonomy, movement, 56–57

Balance, maintaining, 152–154
Battery power, 89
Behavior adaptation, 176
Behavior mode, 178
Behavior selection, 177
Behavior-based robotics, 173, 175
Bipartite, 51–53
Brushed DC motor, 63
Brushless DC motor, 63
Bus, communication, 85

Camera sensor, 92
 for docking, 140
Capek, Josef, 9
Catoms, 69, 78, 79, 84
 energy, 89
CEBOT, 12
Cellular robot, 11–13
Centralized communication, 85
Central pattern generators, 168
Chain-based motion, 57–60

Chain-type, 4, 42
Cheapness, 8, 33–35
Chobie, 20, 48, 61, 78, 79, 84
ckBot, 2, 34
Claytronics, 24
Cluster-flow locomotion, 47, 146–147
Collective actuation, 169
Communication Infrastructure, 83–84
Computing Infrastructure, 83–84
Complexity, 36–37
Configuration representation, 134–138
Connectivity, 100–101, 138
Connectors, 64–74
 desirable characteristics, 65
 mechanical, 69–71
 magnetic, 67–69
 electrostatic, 71–74
CONRO, 15, 16, 42, 78, 79, 84
 connector of, 69–70
 docking, 56
 locomotion of, 43, 166, 167
 self-reconfiguration, 7
Constrained optimization, 169
Constructor robot, self-reconfiguration using, 76
Contingency challenge, 1
Contraction (optimization algorithm), 120
Controller-limited set of configurations, 107
Controllers, role of, 27–29
Cooperative actuation, 169
Coordinate attractor, 130
CPG, 168
Crystalline robot, 15, 19, 46, 78, 79, 84

Design goals, 29–36
Distance metrics. *See* Metrics
Distributed actuator array, 169
Distributed search, 123
Docking, 139–140
Dynamically reconfigurable, 7–8

Environment, 27–29
Electrostatic connectors, 71, 75

# ROBERT J. DIXSON

# MODERN SHORT STORIES IN

# ENGLISH

A NEW REVISED EDITION

Regents Publishing Company, Inc.

Cover design: Paul Gamarello
Text design: Suzanne Bennett & Associates
Illustrations: Connie Maltese

ISBN 0-88345-539-0

Published by Regents Publishing Company, Inc.
2 Park Avenue
New York, New York 10016

Printed in the United States of America

10  9  8  7  6  5  4  3  2  1

# Preface

The stories in this book deal almost entirely with present-day life in the United States. They should be of particular interest to students who, while studying English, also want to learn something of the North American scene.

The stories fall within the form known as the short-short story. Stories of this type provide a convenient study medium, as each is only four or five pages long and can be studied easily within one or two class sessions.

Comprehension questions follow each story. These ten questions, and any others which a teacher may supplement them with, immediately test whether the students have a basic understanding of the story. Teachers should pay close attention to vocabulary, since not all students will understand all the terms used. The exercises should be written and may be supplemented at the teacher's discretion. Generally, the exercises use terms and structures from the story, so teachers have an additional opportunity to check understanding of vocabulary and grammar.

The discussion questions are new in this edition. Teachers

may use them to stimulate the students to use the vocabulary and structures from the story conversationally or for written work. A question such as "Why didn't Mr. Whitney call the police?" can be supplemented with "What did Mr. Whitney do instead of calling the police?" or "What would you have done in Mr. Whitney's situation?" Each question can thus lead to other questions and to interesting and lively class discussions.

*Modern Short Stories in English* is one of a series of three readers for students of English as a second language. The first, *Elementary Reader in English,* is a book of simple reading selections for the beginning student. *Easy Reading Selections in English* is for use on the intermediate level, and *Modern Short Stories* is for the advanced student. If the material in it proves to be too difficult for some classes, it is suggested that one of the two easier readers be tried. For a grammar supplement to this book, the author's *Graded Exercises in English* is suggested.

# Contents

1   My Best Friend
*Karen Tucker White*   1

2   To Love and to Honor
*Octavus Roy Cohen*   7

3   New Applications
*Chandlee Stokes*   12

4   Art for Heart's Sake
*Rube Goldberg*   18

5   That Restless Feeling
*Peg Grady*   24

6   Last Call
*Walter Davenport*   30

7   Red Balloons
*Elmer Davis*   36

8   Decision
*Roy Hilligoss*   42

9   Ten Steps
*Robert Little*   48

10   The Wrong House
*James N. Young*   54

11   Irene's Sister
*Vina Delmar*   60

12   Detour to Romance
*Gilbert Wright*   66

13   Final Break
*Ian S. Thompson*   72

14   A Case of Suspicion
*Ed Wallace*   78

15   Better Late
*Edward Stevenson*   84

# Unit 1: My Best Friend

*Karen Tucker White*

From the time I was a young girl, I had problems with friends. All my girlfriends had "best friends," but I didn't. I never had one special person who always walked home from school with me or called me late at night to talk about things like music or clothes (or later, boys).

My parents used to say that I was a loner, a person who chose her friends carefully, who felt most comfortable when she was alone. It wasn't true. I was never comfortable being alone. I always wished that I had a close friend, like the other kids in my class. Janet Mullaney and Anne Kozach were close friends all through the sixth and seventh grades. How I envied them!

By the time I got to high school, I really began to worry. I knew that I wasn't stupid or ugly or clumsy or any of the other things that kids made fun of, but I still didn't have a special friend that I could share my secret thoughts with.

Occasionally I walked to school with the boy who lived next door. He was my age, so we were in the same grade in school. When we were little, I thought he was gross. I guess that's what all little girls think of all little boys when they're young. Boys are gross, dumb, and dirty. It makes me laugh to think that that's what we thought, but it's true. Doug and I walked to school together, and when he didn't have band practice, we also walked home together, but he didn't count as a friend. He couldn't take the place of a girl my age who felt all the same things I was feeling.

The fact that I didn't have any close girlfriends also seemed to influence my relationships with boys. When I reached my third year of high school, my junior year, all the other girls in the school began to date boys—all except me, that is. I was still a loner.

Now the problem seemed worse. Janet was best friends with a girl named Diane, whom she talked to on the phone for hours. They talked about the boys they liked, they gossiped, they did homework together. Anne had a steady boyfriend, but she had a lot of girls as friends, too, whom she could talk to about all the problems of being sixteen years old. I still had only Doug Thomas, the boy next door. And that didn't count.

I think of my best friend problem as a mountain that I began climbing when I was young. It was a long, hard climb, and I seemed to reach the top of it near the end of my junior year.

The time came for the Junior Prom and I didn't have a date. All of my friends—I guess I should call them acquaintances—were going to this important dance. Of course, they were all talking to their best friends about it. But not only didn't I have anyone to go with, I didn't even have anyone to talk to about it.

My mother told me that she hadn't gone to her Junior Prom either, but I'm not sure she was telling the truth. I think she told me that so I wouldn't feel bad. I felt bad anyway.

I felt so bad that I told Doug about it the next chance I got. He said that he didn't have a date either, and that if neither of us found someone to go with, we could always go to the prom with each other.

2

And that's what happened. Doug and I went to the Junior Prom together. I went because I was embarrassed to stay home, not because I liked Doug. Doug went with me because the girl he liked was already going with someone else. Doug didn't act depressed, and the fact was that we had a pretty good time.

We were invited to a couple of parties after the dance, and then later that week a few acquaintances asked me if I'd like to come to a party they were having. They made it obvious that they wanted me to bring Doug.

I had never paid too much attention to sports in high school, so I didn't know that Doug was now a star swimmer on the school's team. I asked him to go to the party with me and he accepted.

We had a great time at the party, as people crowded around us or joked with us or made references to future events which they hoped we would attend. The attention took me by surprise, but I loved it, since I had never had so many people who wanted to talk to me before.

The new phenomenon continued during the following weeks. Janet called me one day and asked if I wanted to go shopping with her. We had an exhausting yet exhilarating day at the mall and when we got back, she asked me to spend the night at her house. We sat up until 2 A.M. talking about all sorts of things.

A few weeks later, Anne and I worked on a school project together, and I soon realized that I had a couple of "best friends." I was ecstatic.

In the meantime, I went out with Doug a few times. Neither of us was romantically interested in the other, but the other kids in school assumed that we were serious about each other. I liked him, but I was much more interested in my new girlfriends.

Over the summer vacation that year, several events occurred which shook me and my newly found confidence.

Janet and her family moved to another state. We had become good friends by then, and we promised to write and to stay in touch, but we both knew that we probably would never see each other again.

Anne dropped out of school and married her steady boyfriend. She got a job at a local department store, so I see her from time to time; but our interests are different now, so we don't call each other too often.

Doug went to summer swimming camp and fell in love with a girl there. She was a swimmer, too, so they had a lot in common. He dates her now, instead of me, but it's OK. I like her, and they make a pretty good couple.

I'm in my senior year, and my problem of having a best friend doesn't seem so insurmountable to me. I have several girlfriends in whom I confide my secrets. Sometimes I visit them, and sometimes they visit me. It all seems so easy and natural that I wonder what I was so worried about.

When I decided to write this, I analyzed all the events and all my thoughts and came to understand something important. I needed help in being able to get close to people. The help came because Doug was popular. It's an odd way for someone to find friends, perhaps, but at least it worked.

I learned something else, too. All the time I was looking for a best friend, I already had one—Doug. About once a week now, without any special plan, we find a way to walk to school together, as we did in the old days. We talk about all the things that are on our minds, complain about our problems, wonder about the future, remember the past. In other words, we're best friends.

*Comprehension*

1. Why did the narrator of the story want a "best friend"?
2. How did her parents respond to her difficulties?
3. Whom did she walk to school with in her first years of high school? What was her relationship with this person?
4. During her third year of high school, what happened to the narrator's friendships with Janet and Anne?
5. What was her problem at the time of the Junior Prom? How was it solved?
6. Why did people invite her to parties during the weeks following the dance?
7. When did Janet and Anne become her friends? How did it happen?
8. What happened during the summer between the narrator's junior and senior years?
9. At what point in her life is the story being written? What is her attitude toward friendship?
10. What is her relationship with Doug like at the end of the story?

*Exercises*

**A.** Use each of the following terms in a sentence:
best friend, loner, to envy, clumsy, gross, band, to reach something or someplace, steady boyfriend or girlfriend, to make reference to, to stay in touch, to fall in love, to confide in, in other words.

**B.** Match the term in the left column with its OPPOSITE in the right column.

*Example:*   c   **6.** clumsy   **c.** graceful

____ **1.** close        **a.** stranger
____ **2.** stupid       **b.** obscure
____ **3.** ecstatic     **c.** graceful
____ **4.** acquaintance **d.** easy
____ **5.** obvious      **e.** past
____ **6.** clumsy       **f.** distant
____ **7.** tell the truth **g.** smart
____ **8.** future       **h.** lie
____ **9.** from time to time **i.** depressed
____ **10.** insurmountable **j.** often

**C.** The prefix *re-* with verbs means *again*.

He didn't do his work correctly, so he had to do it again.
He had to *redo* it.

Add the prefix *re-* to each verb. Then use the word in a sentence.

*Example:* redesign    The car didn't work well, so the company had to *redesign* it.

  **1.** design      **6.** tell
  **2.** model       **7.** fill
  **3.** arrange     **8.** settle
  **4.** appear      **9.** build
  **5.** consider    **10.** write

*Discussion*

**1.** Do you think it's a good idea for people to have "best friends"?
**2.** Do you have (or have you had) a best friend? What do (did) you talk about with this person?

3. Name some of the ways in which people become friends. How have you made friends throughout your life?
4. What was the difference, for the narrator of the story, between having girls as friends and having boys as friends? What is the difference for you?
5. Do you think it is easier for sports stars to make friends than for other people? Why? What makes a person popular?

# Unit 2: To Love and to Honor

*Octavus Roy Cohen*

It was rather surprising to discover a deep vein of sentiment in little George Potter. I had been his friend and his lawyer for many years and had watched the always fat and once alert little man settle into a domestic routine. He had been moderately successful in business, sufficiently successful to permit him to retire from it and to travel about the world a little if he had wanted to do so. But instead, he and his present wife, Esther, were content to sit night after night in their pleasant and comfortable living room. She kept busy with her hobby of refinishing old furniture, while he passed the time reading or working on his excellent collection of postage stamps.

Looking back over the years of my friendship with Potter, I can see that the vein of romance had probably been there all the time. There was, for instance, his very romantic love affair with Althea Deane—an affair which almost became a scandal. But just when people began to gossip about them, George married her.

That marriage appeared to extinguish George Potter's last spark of romanticism. It never had a chance to be successful, and when Althea left him suddenly, George's friends thought that he was fortunate to lose her. Later came the news of Althea's death while living abroad, and a couple of years later George began to call upon Esther seriously. The people of our group were only slightly interested; it is difficult to become greatly excited over a possible marriage when both the man and the woman are equally dull and uninteresting.

The marriage was a very nice affair. There followed the usual series of parties for the newly married couple. Then it seemed that George and Esther retired from life. His business affairs ran so well that there was little need on George's part for my services as his lawyer, and while I never ceased to like him, we found less and less in common as the years passed. I couldn't imagine that they were happy; perhaps they were contented, but not really happy. There wasn't enough sentiment; that's the way I figured George. And nothing happened to change my opinion until a few weeks before their twenty-fifth anniversary.

It was then that George came into my office, his fat little face shining with enthusiasm, and he told me of his unusual plan for their silver anniversary. His bright little eyes shone as he explained the thing, and I'll confess that I was pretty well confused; not only because his plan was very sentimental and profoundly impressive, but mainly because it was quiet, dull, old George Potter who was planning this thing—the very George Potter who had lived a quiet life since his second marriage and who had avoided social contacts.

According to what George told me, he was doing this thing for Esther's sake. "It'll please her," he explained. "She likes that sort of thing, you know, and this seems to me a real idea. You have to be a part of it, because you were the best man when Esther and I were married. It's just a gesture on my part—a sort of sacrifice to please her."

I'll say this for George; he didn't do things halfway. Instead of the usual party, he presented a perfect duplication of his marriage to Esther twenty-five years before. There was even the same minister—very old now—and the same violinist who had played "Oh, Promise Me" at the other ceremony. A good many of the original guests were there, most of us rather gray-haired now. But the thing was very impressive: Esther in the same bridal dress she had worn twenty-five years before—let out around the hips, perhaps—and carrying a bouquet of bridal roses; the bridesmaids in pink, with bouquets of Killarney roses; even a person to carry the ring. It was great fun and very impressive, whereas one might have expected it to be absurd.

As for Esther, I never saw a woman look more beautiful. She took on an aura of genuine beauty. Of course, she would have been less than human had she failed to respond to this magnificent exhibition of husbandly devotion. George himself was as frightened as he had been on the occasion of their first wedding.

But finally the ceremony was finished, and the guests went to the dining room for the rich supper which had been prepared by special cooks employed for this occasion. George and I were left alone and he sank, exhausted, into a chair. I placed my hand on his shoulder and congratulated him on the success of his party.

"Do you really think it was a success?" he asked hesitatingly. I noticed some wariness in his eyes.

"It was wonderful!" I responded, and I added jokingly, "And you certainly should feel completely married." I expected a short laugh in return, but received none.

"Yes, I do." He became silent for a moment or two. When he spoke again, his tone was deeply serious. "There's something I want to explain to you both as my friend and as my lawyer." He stopped for a moment and then looked up with a curious expression on his face. "You remember my first wife?"

"Althea?" I was surprised by the question. "Yes, of course."

His voice was strange. "Did you know that she died only last year?"

"Good Lord! Are you certain? I thought she died twenty-seven years ago."

"So did I," he said quietly, looking at me long and pensively. "So did I. I thought I was a widower when I married Esther. I only recently discovered that I wasn't. Don't ask me for details, I don't know any. All I know is that Althea didn't die until August of last year. As far as I know, there are no legal complications, but I wanted you to know in case anything ever comes up. I want you to understand that the affair you attended today was a real wedding for Esther and me."

*Comprehension*

1. What was George Potter like? What was his relationship with the narrator?
2. What did George and Esther do to pass the time?
3. What was George's relationship with Althea Deane? How did it end?
4. Why weren't George's friends very interested in his relationship with Esther?
5. How did the narrator feel about George over the years?
6. What plan did George make for his twenty-fifth wedding anniversary? Why did this seem unusual to his lawyer?
7. Describe the ceremony. What was special about the minister and the violinist?
8. How did George and Esther react to the ceremony?
9. What was George's manner as he told his friend the real reason for the wedding ceremony?
10. What was that reason?

*Exercises*

A. Use each of the following terms in a sentence:
rather, to do so, to keep busy, to pass the time, looking back, romance, scandal, to become excited over, in common, to do something halfway, impressive, as for, wariness, complication, in case.

B. Circle the term in parentheses which correctly completes the sentence.

*Example:* One generally buys stamps in a (bookstore/ⓟost office/movie theater/grocery store).

1. If you were involved in a scandal, you would probably be (happy/in a hurry/embarrassed/sleepy).
2. A bouquet is made of (stones/groceries/stamps/flowers).
3. They were talking about other people. They were (leaving/planning/gossiping/settling down).
4. I never ceased to like him. I never (started/avoided/continued/stopped) liking him.

5. There was some wariness in his eyes. He was being (cautious/ecstatic/impressive/ugly).
6. A domestic routine is one which takes place mostly at (the office/home/the swimming pool/a restaurant).
7. A person whose face is shining with enthusiasm is probably (sad/happy/sick/exhausted).
8. A curious expression is a (handsome/shy/dull/strange) expression.
9. If something comes up, it (dies/avoids/stops/arises).
10. To change one's opinion is to change one's (clothes/furniture/ideas/services).

C. The endings *-or* and *-er* are used with some verbs to form nouns, indicating a person or an agent that performs the action of the original verb.

> She illustrates magazines.
> She is an *illustrator* who draws for magazines.
>
> He waits on tables in that restaurant.
> He's a *waiter*.

Change the following verbs to nouns by adding *-or* or *-er*. Then use each word in a sentence.

*Example:* painter    He went to art school and became a *painter*.

1. paint
2. act
3. sing
4. govern
5. employ
6. collect
7. educate
8. advise
9. instruct
10. think

*Discussion*

1. How does your (or your parents') life style compare to that of George and Esther Potter?
2. What do you think the narrator meant when he referred to George's *spark of romanticism?* Do you think of yourself as a romantic person?
3. What legal complications might there have been had Althea returned from abroad after George married Esther?
4. What are wedding ceremonies like in your family?
5. What do you think was going through George's mind as he "remarried"?

# Unit 3: New Applications

*Chandlee Stokes*

Miriam Storley left the bank at 4:15 exactly. People along Division Street said you could set your watch by Miriam; she always left her job at the First State Bank of Cannon Falls at this hour, Monday through Friday, except on holidays. On Fridays she returned to work the six-to-eight P.M. shift. On this particular day, a Monday, she stopped after closing the front door to the bank in order to look at the window display.

Miriam had spent the better part of the afternoon arranging gift items in the bank's window. First State, which is how everyone in town referred to the bank, was having a promotion in order to attract new business. They were offering gifts which ranged in value all the way from a pocket calculator to a color TV. The value of a new depositor's gift depended on how much was initially deposited.

The display in the window was attractive, but Miriam wondered where the new business was going to come from. Cannon Falls wasn't a one-stoplight town, but it wasn't a great metropolis either. There just weren't that many people to warrant an extravagant new business promotion such as this. The bank manager, Al Gropin, had even invested in some full-page advertisements in the local paper and had hired some clowns to perform on the street in front of the bank—all to try to attract new customers.

But Miriam didn't linger long in front of the window, and she didn't waste much time on her thoughts of Al's grand schemes. Her mission today was the same as it had been every weekday for the past several weeks.

She nodded at passersby, shopkeepers, and neighbors as she walked purposefully along the wide sidewalk toward The Computer Shack. There was a pleasant expression on her face as she smiled and said her "hellos" and "good afternoons" and "how are yous" to the people she saw almost every day of her life. Her daily meeting with Officer Quanbeck never failed to amuse her. She smiled to herself as they exchanged greetings and wondered whether he would feel as stupid as he looked after she pulled off the crime of the century.

"Right on time, as usual, eh, Mrs. Storley?" The thin, kindly looking man behind the counter in The Computer Shack seemed to have a perpetual smile on his face. Every day for the past several weeks, Tobe Barksdale had a short, simple conversation with this woman from the bank down the street. She said she wanted to buy the home computer which he had hooked up to a printer and which was fully operational, but so far all she did was sit and play with it.

Tobe didn't mind the intrusion, though. Even though he opened his shop, gleamingly filled with electronic toys and machines, at noon, the majority of his customers came after six P.M. At first, he had closed the store at eight, but the numbers of people interested in the latest gadgetry forced him to stay open later and later, and now he wasn't closing until ten o'clock.

He could have insisted that his daily visitor make up her mind

**13**

about the computer, or at least stop using the same program all the time, but she wasn't really any bother, and lately she had acquired such a solid knowledge of the field that he actually enjoyed her increasingly complex questions. She challenged his imagination, probing to see just how far a computer could go, just how much a simple machine could do.

Tobe probably knew as much about computer hardware and software as anybody in the entire town of Cannon Falls. Hardware and software. These were terms the general public rarely heard when Tobe began working a number of years back. Now, everyone used the terms to refer to the computers themselves and the programs which told the machines and operators what to do.

Miriam Storley had a long way to go to catch up with Tobe in her knowledge of this complex field, but she seemed determined, and Tobe was a patient instructor. Each day she would come to him with a new type of problem, an unusual twist, a tricky flow of information or instructions which she wanted to master. Every day he would guide her through the intricacies of the model which was advertised as the "latest, most technologically advanced home computer ever designed." Every day she would listen and absorb, and then experiment for herself. She brought her own tapes and never seemed to tire of learning, even after a day's work. Tobe believed in leaving people to themselves, so when the lesson was over and Miriam sat at the console, enwrapped in her task at hand, he busied himself in another part of the store.

Miriam's teen-age son, who liked to be called by the nickname Zee, had introduced her to the world of computers through his interest in video games. True, she dealt with computers at the bank every day in her job, but somehow they were just a part of the bank; they didn't touch her.

She learned from her son and, almost by accident—as most great discoveries in the world seem to be—she discovered that the latest version of the home-type computers was actually compatible with the one she worked with in her office at First State.

The idea came to her at the end of a particularly tiring day as she tallied the day's receipts and entered them into her desk-top computer. It was foolproof! She could transfer funds from various accounts which were relatively inactive by tampering with the program. If she did it skillfully enough, she would never be caught. She would set up some fictitious accounts in other banks in the state, transfer funds, disguise herself and go to the other banks in order to withdraw the money, and then return the program to its original condition. No one would ever be able to figure out what she had done or where the money had gone. And even if they did trace it, they would never suspect her. How could they?

She decided not to risk working on the program she needed at home, since Zee might see what she was doing. Tobe Barksdale's shop was the perfect cover, and that pleasant man certainly wouldn't suspect her. He didn't even seem to mind letting her use his floor-model computer.

After months of preparation, Miriam carried out her plan. She called Mr. Gropin to say that she was ill and couldn't come to work. Then she drove to Mankato and Red Wing, disguised, and picked up her money. All went well until she arrived home to find Officer Quanbeck and several others waiting in her living room to arrest her for fraud and bank robbery.

As a kindness, to assuage her curiosity, Tobe Barksdale was there, too. He explained, "Your plan was brilliant, Miriam, and you were an excellent student. Indeed, I taught you almost everything you know. But I didn't teach you everything I know. The computer you worked on recorded everything you did on a master tape which I observed every afternoon after you left. After all, I had to see what kind of progress my pupil was making, didn't I?"

*Comprehension*

1. When did Miriam Storley finish work every day? What did people say about her punctuality?
2. Why had she arranged the bank's window in a special display?
3. Who was the bank manager? What had he done to promote new business?
4. Who was Officer Quanbeck? When did Miriam see him?
5. What did Miriam do at The Computer Shack every day?
6. Who was Tobe Barksdale? What were his store's hours of operation?
7. How did Miriam get interested in computers?

**15**

8. Why did she want to tamper with the bank's program?
9. Where did she go on the day that she called in sick? Why did she need a disguise?
10. How did she get caught?

*Exercises*

A. Use each of the following terms in a sentence:
to set one's watch by, shift, promotion, extravagant, to waste time, grand scheme, as usual, to hook up, so far, at first, gadgetry, just how far, hardware, software, nickname, by accident, to figure out, curiosity, after all.

B. Many words in English may be used as a verb or noun without any change in form.

> The publisher will *print* several thousand copies.
> The *print* on this page is small and black.

Use each of the following words in a sentence, first as a verb and then as a noun.

1. bank
2. shift
3. display
4. waste
5. smile

6. transfer
7. master
8. touch
9. experiment
10. disguise

C. Match the term in the left column with one which is SIMILAR in the right column.

*Example:*  <u>e</u> 10. shortened name      e. nickname

___ 1. conversation     a. take in
___ 2. disguise         b. make happen
___ 3. absorb           c. add
___ 4. foolproof        d. talk
___ 5. touch            e. nickname
___ 6. pull off         f. cover
___ 7. value            g. test
___ 8. tally            h. worth
___ 9. probe            i. affect
___ 10. shortened name  j. infallible

*Discussion*

1. Would you like to own a home computer? How would you use it?

2. What jobs or functions can today's home computers accomplish?
3. Do you ever play any video games? Which ones? Do you think they have any value?
4. What do you think the next generation of computers will enable us to do?
5. Do you think that in a real-life situation, Miriam would have been caught? What other computer-related crimes are possible?

# Unit 4: Art for Heart's Sake

*Rube Goldberg*

Keith Koppel, private duty nurse to the extraordinarily wealthy Collis P. Ellsworth, was glad to leave his patient's room to answer the door. He had had a tiring morning trying to get Ellsworth to cooperate in his own recovery. As soon as Koppel discovered that the caller was Ellsworth's doctor, he began to complain.

"I can't do a thing with him," he told Dr. Caswell. "He won't take his juice. He doesn't want me to read to him. He hates listening to the radio or watching TV. He doesn't like anything."

Actually, he did like something: his business. The problem was that while he was still a fabulously wealthy man, he had recently begun to make big mistakes. He insisted on buying companies at very high prices, only to watch them fail or go bankrupt.

Ellsworth was in pretty good shape for a 76-year-old, but his business failures were ruinous to his health. He had suffered his last heart attack after his disastrous purchase of a small railroad in Iowa. The health problem he suffered before that came about because of excitement over the failure of a chain of grocery stores, stores which he had purchased at an inflated price. It seemed that all his recent purchases had to be liquidated at a great sacrifice to both his pocketbook and his health. They were beginning to have serious effects.

Dr. Caswell had done his homework, however. He realized that he needed to interest the old man in something which would take his mind off his problem and redirect his energies. His answer was art. The doctor entered his patient's room.

"I hear that you haven't been obeying orders," the doctor said.

"Who's giving me orders at my time of life?"

The doctor drew up his chair and sat down close to the old man. "I've got a suggestion for you," he said quietly.

Old Ellsworth looked suspiciously over his eyeglasses. "What is it, more medicine, more automobile rides, more foolishness to keep me away from my office?"

"How would you like to take up art?" The doctor had his stethoscope ready in case the suddenness of the suggestion proved too much for the patient's heart.

But the old man's answer was a strong "Foolishness!"

"I don't mean seriously," said the doctor, relieved that nothing had happened. "Just play around with chalk and crayons. It'll be fun."

But after several more scowls, which were met with gentle persuasion by the wise doctor, Ellsworth gave in. He would, at least, try it for a while.

Caswell went to his friend Judson Livingston, head of the Atlantic Art Institute, and explained the situation. Livingston produced Frank Swain. Swain was an 18-year-old art student, quite good, who needed money to continue his education. He would tutor Ellsworth one afternoon a week for ten dollars an hour.

Their first lesson was the next afternoon. It was less than an overwhelming success. Swain began by arranging some paper and crayons on the table.

"Let's try to draw that vase over there," he suggested.

"What for? It's only a bowl with some blue stains on it. Or are they green?"

"Try it, Mr. Ellsworth, please."

"Umph!" The old man took a piece of crayon in a shaky hand and drew several lines. He drew several more and then connected these crudely. "There it is, young man," he said with a tone of satisfaction. "Such foolishness!"

Frank Swain was patient. He needed the ten dollars. "If you want to draw, you will have to look at what you're drawing, sir."

Ellsworth looked. "Gosh, it's rather pretty. I never noticed it before."

Koppel came in with the announcement that his patient had done enough for the first lesson.

"Oh, it's pineapple juice again," Ellsworth said. Swain left, not sure if he would be invited back.

When the art student came the following week, there was a drawing on the table that had a slight resemblance to a vase. The wrinkles deepened at the corners of the old gentleman's eyes as he asked, "Well, what do you think of it?"

"Not bad, sir," answered Swain. "But it's not quite straight."

"Gosh," old Ellsworth smiled, "I see. The halves don't match." He added a few lines with a shaking hand and colored the open spaces blue, like a child playing with a picture book. Then he looked towards the door. "Listen, young man," he whispered, "I want to ask you something before old Pineapple Juice comes back."

"Yes, sir," answered Swain politely.

"I was thinking—do you have the time to come twice a week, or perhaps three times?"

As the weeks went by, Swain's visits grew more frequent. When Dr. Caswell called, Ellsworth would talk about the graceful lines of the chimney or the rich variety of color in a bowl of fruit.

The treatment was working perfectly. No more trips downtown to his office for the purpose of buying some business that was to fail later. No more crazy financial plans to try the strength of his tired old heart. Art was a complete cure for him.

The doctor thought it safe to allow Ellsworth to visit the Metropolitan Museum, the Museum of Modern Art, and other exhibitions with Swain. An entirely new world opened up its mysteries to him. The old man showed a tremendous curiosity in the art galleries and in the painters who exhibited in them. How were the galleries run? Who selected the pictures for the exhibitions? An idea was forming in his brain.

When the late spring began to cover the fields and gardens with color, Ellsworth painted a simply horrible picture which he called "Trees Dressed in White." Then he made a surprising announcement. He was going to exhibit the picture in the summer show at the Lathrop Gallery.

The summer show at the Lathrop Gallery was the biggest art exhibition of the year—in quality, if not in size. The lifetime dream of every important artist in the United States was a prize from this exhibition. Among the paintings of this distinguished group of artists, Ellsworth was now going to place his "Trees Dressed in White," which resembled a handful of salad dressing thrown violently against the side of a house.

"If the newspapers hear about this, everyone in town will be laughing at Mr. Ellsworth. We've got to stop him," said Koppel.

"No," warned the doctor. "We can't interfere with him now and take a chance of ruining all the good work which we have done."

To the complete surprise of all three—and especially Swain —"Trees Dressed in White" was accepted for the Lathrop show. Not only was Mr. Ellsworth crazy, thought Koppel, but the Lathrop Gallery was crazy, too.

Fortunately, the painting was hung in an inconspicuous place, where it did not draw any special notice or comment.

During the course of the exhibition, the old man kept on taking lessons, seldom mentioning his picture. He was unusually cheerful. Every time Swain entered the room, he found Ellsworth laughing to himself. Maybe Koppel was right. The old man was crazy. But it seemed equally strange that the Lathrop committee should encourage his craziness by accepting his picture.

Two days before the close of the exhibition, a special messenger brought a long, official-looking envelope to Mr. Ellsworth while Swain, Koppel, and the doctor were in the room. "Read it to me," said the old man. "My eyes are tired from painting."

It gives the Lathrop Gallery great pleasure to announce that the First Prize of $1,000 has been awarded to Collis P. Ellsworth for his painting "Trees Dressed in White."

Swain and Koppel were so surprised that they could not say a word. Dr. Caswell, exercising his professional self-control with a supreme effort, said, "Congratulations, Mr. Ellsworth. Fine, fine. . . . Of course, I didn't expect such great news. But, but— well, now, you'll have to admit that art is much more satisfying than business."

"Art has nothing to do with it," said the old man sharply. "I bought the Lathrop Gallery last month."

*Comprehension*

1. Who was Keith Koppel? What was his complaint about his patient?
2. What had Collis P. Ellsworth begun doing recently that had his doctor concerned?
3. What did Dr. Caswell prescribe in order to help his patient avoid further health problems?
4. Who was Frank Swain? How did he get the job as tutor?
5. What was the difference between the first Ellsworth drawing of the vase and the second drawing?
6. How often did Ellsworth want Swain to come?
7. What happened when Ellsworth began to visit museums and galleries?
8. What was "Trees Dressed in White"? How good was it?
9. What was the Lathrop Gallery? Why did some people think the Lathrop committee was crazy?
10. What was the final twist to the story?

*Exercises*

A. Use each of the following terms in a sentence:
as soon as, to go bankrupt, chain of stores, to be in good shape, to take one's mind off something, to keep away from, in case, to go by, not bad, a handful, salad dressing, to take a chance, official-looking, of course, nothing to do with it.

**B.** Match the term in the left column with one which has the SAME meaning in the right column.

*Example:*   i   2. dull     **i.** uninteresting

| | | |
|---|---|---|
| ___ | **1.** wealthy | **a.** unhappy |
| ___ | **2.** dull | **b.** ended |
| ___ | **3.** crazy | **c.** continue |
| ___ | **4.** keep on | **d.** unusually |
| ___ | **5.** chance | **e.** opportunity |
| ___ | **6.** extraordinarily | **f.** insane |
| ___ | **7.** over | **g.** bear |
| ___ | **8.** sad | **h.** prize |
| ___ | **9.** stand | **i.** uninteresting |
| ___ | **10.** award | **j.** rich |

**C.** The prefixes *un-*, *in-*, *im-*, *il-*, and *dis-* are used with adjectives and adverbs to give a negative or opposite meaning.

Give the negative form of each of the following adjectives or adverbs. Then use the word in a sentence.

*Example:* uninteresting     This book is too uninteresting to finish.

| | |
|---|---|
| **1.** interesting | **6.** loyal |
| **2.** patiently | **7.** honest |
| **3.** agreeable | **8.** sure |
| **4.** able | **9.** attentive |
| **5.** legal | **10.** politely |

*Discussion*

1. In what other ways could a person in Mr. Ellsworth's condition have redirected his energies?
2. Have you ever produced any artwork? What kind?
3. What are some ways in which people of all ages can stay in shape? What do you do to get or stay in shape?
4. Do you enjoy going to art museums? Which ones have you visited?
5. What is the purpose of an art exhibition?

# Unit 5: That Restless Feeling

*Peg Grady*

Mrs. Feldmeyer found her in her bedroom crying. She stood in the doorway and said seriously, "I came over to borrow your vacuum cleaner, Carolyn. Mine's broken. The door was open, so I just walked in. What on earth is the matter?"

Carolyn sat up and dried her eyes. She was embarrassed that Mrs. Feldmeyer had come in while she was crying. She had been caught letting her emotions get the better of her. She shook her head to control herself. "Nothing. Really. I'm OK. Of course you may borrow the vacuum. It's in the closet. I'll get it." She rose shakily.

Mrs. Feldmeyer put her hands on her hips, cocked her head,

and looked skeptical. "Nothing, eh? It's because Joe wants to leave Libertyville, isn't it? Of course it is."

Carolyn pushed her thinning blond hair out of her eyes and said, somewhat defiantly, "I won't do it. I won't."

"He's got that restless feeling, hasn't he? I've known that boy since he was a little tyke. I guess a person could see it coming. This is a small town, and a fellow like Joe Conly gets restless and needs to get out into the world. I don't think you're going to have too much luck trying to fight an urge like that one, my dear."

"But I don't have the urge. I won't move from town to town all my life. I like it here. I know everyone. It's peaceful. I don't have to worry about things. Well, I do worry about some things, but you know what I mean. It's like an old pair of slippers. They're comfortable. Why would I want to buy a new pair?"

"Carolyn, you married Joe for better or worse. He's got something inside him that says, 'Get out there on the road.' Now how are you going to deal with that?"

"I don't know," she said, her head bowed slightly.

"Of course you don't. You're just a child. You'll fight it and fight it and the next thing you know, twenty years will have passed, and by then it will be too late to do any good. Unless you find someone to tell you what to do to begin with. Someone who knows."

Carolyn was still depressed. "Who could know the answer, Mrs. Feldmeyer? Who could tell me what to do? I've argued and reasoned with Joe until I'm almost crazy, but he doesn't even listen anymore. His mind is made up. He says that I could find another job easily with my talents, and that this is something he needs to do."

"Is he right? About you finding another job easily, that is?"

"I suppose so. That's not the part I'm upset about. I'd like to take a few years off from work anyway. You know, when we have children. I've always thought that it would be great to stay home and be with the children for a few years. Until they're in school. It's just that I've never wanted to travel. I always thought that we'd live a quiet, uncomplicated life right here in dull little Libertyville, but Joe seems so restless."

"Like Mr. Feldmeyer."

Carolyn looked up, surprised. "You mean Mr. Feldmeyer used to . . . wanted to . . . ?"

"He was the hardest man to hold in this town. He was aching to get on the road. Just like your Joe. But he stayed. He stayed all his life, God bless him."

"Why?" Carolyn asked. "What did you do?"

"I always gave in. It worked, too."

Carolyn looked confused and disappointed. "Oh."

"But he didn't know it. Oh, I suppose you could call what I did 'being manipulative,' but we were a happy pair, we were. He got his way and I got mine. 'Course, I don't know if what I did would work with your Joe."

"Well, at least tell me about it. What did you do? What do you mean, you 'gave in'? How could you get your way by giving in?"

"Whenever Mr. Feldmeyer got that restless feeling, I always took him on a trip. Just for a week or two. And I kept him jumping every minute of those trips. I never minded short vacations anyway; they were fun. Well, by the time that man would get home, he'd be so tired of jumping around from place to place, he wouldn't have left Libertyville for a million dollars."

"Oh, I don't think that would be right. It would be like tricking Joe. And besides, it probably wouldn't work."

"Maybe it wouldn't. I'm the last person in the world to try to tell folks how to live their lives, Carolyn, but it seems to me that you're pretty unhappy right now. You're going to have to do something to solve this problem you and Joe have. Why not give it a try. Tell him you'd like to take a short vacation before you change your life style."

They did it. They left Tennessee and headed east, stopping in Louisville and Lexington in Kentucky and then wandering north through Cincinnati, Columbus, and Dayton, Ohio, and then back west through Indianapolis before coming back by way of Nashville. They were gone seven days. Each day they rose at dawn and got on the road, stopping at every historical marker and tourist attraction they saw. Carolyn planned almost every moment, filling each with some purposeful activity. She also filled Joe with fast food and stale coffee, remarking that that was all they had time to eat since their main goal was to see as much as they could. Joe's enthusiasm began to wane on the third day out. A lifeless cloud began to cover his eyes.

Mrs. Feldmeyer dropped over the day after they returned to bring back the vacuum cleaner she had borrowed. Joe was still asleep, but Carolyn was up getting ready for work.

Carolyn's news was "good," but there was an odd quality to her voice as she told it. "About halfway through the trip, Joe stopped talking about going on the road," she related. "He hasn't said a word for days about being restless. He says he's anxious to get back to work at the plant. He's even suggested that we invite his brother and his family to visit us this Christmas instead of going to Chicago to visit them, as we had planned."

"I'll bet he said that home never looked so good to him, too."

"Exactly. I probably won't even be able to get him to go to a movie for a month." Carolyn managed a slight smile.

"You look tired, dear, not too happy, and more than a little weary."

"Weary? Yes, I suppose I am. Weary of this town and this life. I never knew how old and dull and tiresome this town was. I thought we'd spend our entire lives here. I wanted to have a family and have a life for my children just like the life I had. But now . . ."

Mrs. Feldmeyer looked at Carolyn seriously for a moment. "You're just tired from the trip, dear. I'll bet all that traveling was a strain on you. It was such a long trip . . ."

"It wasn't the travel." When Carolyn looked up, her eyes were different. They were shining with a new life and vibrancy she had never felt before. "The trip was exciting. I had a wonderful time!"

*Comprehension*

1. Why did Mrs. Feldmeyer visit Carolyn at the beginning of the story?
2. What was Carolyn doing when Mrs. Feldmeyer arrived? Why?
3. Who was Joe Conly? What did he want to do?
4. Why did Carolyn want to stay in Libertyville?
5. What did she want to do when she had children?
6. What was Mrs. Feldmeyer's advice?

7. Where did Carolyn and Joe go on their short vacation?
8. How did they spend their time on the vacation?
9. What was Joe's reaction to their week on the road?
10. Describe the change that came over Carolyn while on vacation.

*Exercises*

A. Use each of the following terms in a sentence:
what on earth, to be the matter, to get the better of, to be OK, to cock one's head, defiantly, tyke, for better or worse, to deal with, almost crazy, to be upset about something, manipulative, fast food, tourist attraction, on the road.

B. Supply the appropriate tag ending for each of these sentences.

*Example:* You're going, _____?
You're going, aren't you?

1. It's because he wants to leave, _____?
2. He wants to move, _____?
3. You're tired from your trip, _____?
4. I don't have to leave this town, _____?
5. She decided to work for a while, _____?
6. The vacuum cleaner is in the closet, _____?
7. She wasn't tired from her trip, _____?
8. They've already left for Chicago, _____?
9. We can't go on vacation with you this year, _____?
10. You're not crying because you're happy, _____?

C. The opposite of the *-ful* ending for adjectives is *-less*, which has the general meaning of *without*. Change the following nouns to adjectives, first by adding *-ful* and then by adding *-less*. Then use each word in a sentence.

*Example:* restful    We had a *restful* vacation.
restless    I'm too *restless* to take a long car trip.

1. rest
2. care
3. pity
4. grace
5. hope
6. event
7. use
8. thought
9. fear
10. color

28

*Discussion*

1. Do you think you would enjoy living "on the road"? What do you think it would be like?
2. What do you think of the way in which Mrs. Feldmeyer and Carolyn manipulated their husbands? Was it unfair?
3. What are the advantages of living in a small town? The disadvantages?
4. Do you ever get a restless feeling? What do you do about it?
5. Where did you go on your last vacation? Was it rushed? What kind of food did you eat? How did you feel when you returned home?

# Unit 6: Last Call

## *Walter Davenport*

Most people would have told young Sherrill the facts at the beginning, using no more words than were necessary for simple courtesy.

But that wasn't Ham's way.

Ham Mason probably would have been a good public speaker if he had wanted to. He truly loved to talk. Ham was content, however, to run his small hotel in Owensboro. He owned the Mason House, was its general repairman, and could usually be found at its front desk chatting with anyone and everyone who wandered by.

"Glad to have you, Mr. Sherrill. Will you be staying with us a few days?" Sherrill had rented a car and driven in from the airport. He had made his reservation at the Mason House by mail weeks ago.

"A few days, at least, Mr. . . . ?"

"Mason. I'm the proprietor. Everyone 'round here calls me Ham, though. Let me see." Ham looked at the board where, in great disarray, hung the keys to the various rooms. "Four, five, six, se— I can let you have number six. It's got a desk in it, if you think you'll want to do some work. I notice you've got a hefty briefcase there."

Ham talked all the while Mr. Sherrill filled out the registration card. He talked as they walked down the hall to Sherrill's room, and he talked as he opened the curtains in the room, letting in the last rays of the setting sun. Ham instinctively knew that young people were durable about listening to him, whereas an older person might have silenced him minutes ago.

"From New York, you say? What's your business?"

"Investments," said the young businessman. "I'm with Chaffee and Bates, one of the oldest investment houses in New York. I haven't been with them long. I guess that's why they sent me out chasing old business, trying to make new business out of it. I've come to see Mr. Edward Colesberry. Do you know him?"

A true orator takes advantage of any opportunity. Ham saw a real chance to exhibit his gifts of eloquence.

"Ed Colesberry?" he almost shouted, "Why, Mr. Sherrill, there's no one in this whole town who could tell you he was closer to him. So you came all the way from New York City to see Ed Colesberry, eh?" Ham was warming up to his newly found audience and was very pleased with himself.

"Him and a few others," said the young man. "Mr. Colesberry used to do business with us, but he hasn't for a while. The firm decided to send out us new fellows to look up old customers."

"Well, I don't plan to take up too much of your valuable time, young fellow, but if you'd like, I could tell you something about this man you've come so far to see."

Sherrill was eager to hear and learn something about Ed Colesberry, so he joined Ham in the hotel dining room for a cup of coffee while the hotel man, elated with his audience, began his tale.

"Ed Colesberry was seventy years old last month. Getting along in years. He and I went to school together, so you know that I knew him well all my life. Popular boy, Ed was. Smart, too. Half the businessmen in town wanted Ed to work for them when he graduated from high school.

" 'Ham,' Ed said to me the day after graduation, 'I'm going to work for Willis Deemer for a while. Then as soon as I get some money saved, I'm going to New York to do some big things with my life. I'm not going to bury myself here in Owensboro. I'm not going to be happy until I hear the flight announcer over the loudspeaker saying, "Last call for Trans-Global's flight to New York City!" '

"Well, Ed worked in Willis Deemer's hardware store for a year and did so well that Deemer, who was getting ready to retire, offered to make Ed a partner. Ed said no several times to that offer, when all of a sudden, Deemer died. Willis's widow begged Ed to stay on and take over the business for a while because it was all that Willis had left to her.

"Ed stayed a little longer, all the while thinking that he didn't want to be tied down, but soon the Deemer hardware store was the biggest in the county and getting bigger. Meanwhile, Ed fell in love with Maureen Brent, whose father had the area's largest real estate operation.

"Ed and Maureen planned to marry and then move to New York, where he still planned to be 'someone.' He used to tell her that he couldn't wait to hear that announcement: 'Last call for the flight to New York.'

"Well, Sherrill, Maureen got sick, so they postponed the wedding a few times, and when they did get married, her doctor told them they shouldn't move so close to the ocean as New York, at least not for a year or so.

"So they stayed in Owensboro, getting richer at the store but still planning to go to New York at any cost. Next thing you know, Ed and Maureen had a baby and Maureen's father talked Ed into taking over one of his real estate projects. It made them a great deal of money, so they decided to hold onto Ed's dream but to delay it for a few years. It seemed that everything Ed touched turned to gold in those days. The project got bigger and bigger and everyone trusted Ed. Owensboro probably never had a bigger moneymaker than that project and Ed Colesberry.

"Wouldn't you know it? The next thing that happened was that Maureen's father died, and Ed had to take over the entire company. He was madder than a bear, and he had had enough.

He sold half his interest in everything he owned, made arrangements for others to run the businesses, and tried to sell his house.

"They had a second child by then, but he and Maureen were ready to leave for New York at last. 'Ham,' Ed said to me, 'I'm going to New York. Yes, sir. The time has come. Maureen is still somewhat ill, so she'll need special doctors that are only in New York. I want to send the kids to fine schools, too. Yes sir, soon I'll be hearing that announcement: "Last call" that announcer will say.'

"Well, Mr. Sherrill, just at that time Tom Staub, the cashier at the savings bank, ran off with all the money and the bank almost failed completely. The people in town had lost most of their money and they wanted someone to come to their rescue and save the bank. You can guess whom they turned to, can't you? Ed Colesberry.

"Maureen herself wanted to go to New York by that time, but she persuaded Ed to step in even though he never had any banking experience. Of course it took a little time, but Ed saved it.

"He came out of it president of the bank. It's the biggest savings bank in this end of the state today. But when Ed took over that bank he came to me and said that no matter what happened, he was going to New York and nobody could stop him. I couldn't help laughing a little."

Ham paused impressively.

"So I'll find him at the savings bank, then?" said young Sherrill.

"No, you won't. Ed finally made it to New York. I was coming to that part, even though it's taken me a while. Ed died last week. In his will, which he made right after they made him president of the bank, he left a million dollars to the town (with the rest of his sizable estate going to his wife and children) provided that they bury him in New York. He was determined to get there somehow."

*Comprehension*

1. What did Ham Mason do for a living?
2. Who was Sherrill? How did he get to Owensboro?
3. Why had Sherrill's firm sent him there?
4. Why was Ham eager to talk to Sherrill?

5. What was Ham's relationship to Ed Colesberry? How well do you think the two men knew each other? Why?
6. What was Colesberry's dream?
7. Who was Maureen Brent? How many children did she and her husband have?
8. Why did Ed take over the hardware store business? How did the business do?
9. What jobs did Ed have after working at the hardware store? How did he do in each job?
10. How did Ed finally get to New York? How does this relate to the title of the story?

*Exercises*

**A.** Use each of the following terms in a sentence:
public speaker, reservation, hefty, setting sun, to silence someone, to warm up to, a while, last call, all of a sudden, to stay on, to take over, to be tied down, a year or so, to turn to gold, to turn to someone, even though, to step in.

**B.** Circle the term in parentheses which best completes the sentence.

*Example:* A ((vase)/desk/briefcase/envelope) is used to hold flowers.

1. Most men's shoes are tied with (handkerchiefs/laces/vests/boots).
2. Most pants are worn with a (sleeve/cuff/belt/necktie).
3. When a window is shut, it is (open/clean/broken/closed).
4. I park my car in a parking (lot/highway/room/station).
5. My (nose/chin/forehead/collar) is not a part of my face.
6. A strong-willed person has (education/determination/physical strength/a disease).
7. You probably would not be able to buy a (screwdriver/hammer/curtain/box of nails) in a hardware store.
8. A real estate salesperson sells (fruit/cars/beds/houses).
9. A million dollars is $(1,000/1,000,000,000/1,000,000/100).
10. They talked Ed into it; in other words, they (persuaded/narrated/orated/spoke) him.

**C.** Fill in the missing form of the word.

*Example:* <u>curiosity</u> _____ _____
( <u>curiosity</u>    <u>curious</u>    <u>curiously</u> )

| Noun | Adjective | Adverb |
|------|-----------|--------|
| 1. _____ | courteous | _____ |
| 2. contentment | _____ | |
| 3. _____ | _____ | intelligently |
| 4. _____ | | durably |
| 5. _____ | serious | _____ |
| 6. value | _____ | _____ |
| 7. anger | _____ | _____ |
| 8. _____ | eloquent | _____ |
| 9. _____ | _____ | silently |
| 10. warmth | _____ | _____ |

*Discussion*

1. Do you know any people who, like Ham Mason, like to talk a lot? What are they like? Are you ever like that?
2. What is an investment house? What does someone like Sherrill do for a living?
3. What do you imagine was the secret of Ed Colesberry's success?
4. Do you have a will? Do you think it's a good idea to have one? To whom would you leave your possessions?
5. Do you have any dreams similar to Ed's? What are they?

# Unit 7: Red Balloons

*Elmer Davis*

After it happened, Lundy told himself that the temptation had been too great. In fact, he had never been truly tempted before, for the opportunity had never arisen. He had gone to the bank —the branch bank in the poor, rundown neighborhood to which he had recently moved—to review his dwindling investments and savings. He had recently lost a great deal of his financial security in a crazy attempt to make money by playing the stock market so that he could give up his job and live in Florida. He took his safe-deposit box to one of the booths where people shut themselves in while they open their boxes in order to put in or remove valuables. The booth had just been vacated by a fat woman wearing many jewels who had left it covered with torn papers.

A little annoyed, Lundy brushed away the torn papers—and came upon an envelope filled with money which the fat woman had obviously overlooked. The recent recession had frightened many people; the fat woman looked like the sort of person who would turn her bank balance into cash and lock it up in her safe-deposit box. Lundy half opened the door to call her back and saw her walking out of the bank. Quickly he shut the door and counted the money. Nearly ninety thousand dollars; enough to keep a man comfortably in some little Florida town for the rest of his life.

Quickly, Lundy slipped the envelope into his inside pocket.

Then he left the bank, crossing the street into a little park with a high iron fence around it. It was, he knew, a private park, the possession of the old families that had once lived on the square; at night its gates were locked and a watchman guarded it. But by day it was open to all. He sat down on a bench, trembling in the winter wind; the envelope in his pocket felt like a piece of hot metal.

What a fool he had been! He had thought when he took it that it wouldn't be missed for a month—not until the woman came again to get something else. But if she kept all her money in her safe-deposit box she might come back and find it missing tomorrow—or even this afternoon. The bank employees would remember Lundy, as he had recently rented his safe-deposit box; they might remember that he had followed her into the booth. If he gave up his job now and left for Florida, that would be a confession. But tonight, tomorrow, he might be questioned, his rooms examined. Where could he hide the money?

His throat was dry. He got up and walked to the center of the park, where he had seen a drinking fountain. Unable to decide what to do, he stared at the fountain, at its tall concrete base. Then his eyes narrowed; the base was broken on one side and had a hole big enough to put your hand through. Inside, there was a dark space where no one would think of looking for anything; where a man who had hidden something could come back and get it almost anytime.

Beside the drinking fountain, Lundy kneeled down; anyone who had passed would have seen only a man with an unbuttoned overcoat hanging loose about him, kneeling down, tying his shoe. But when he went on, the envelope of money no longer lay like a piece of hot metal in his pocket. He had hidden it in the hole at the base of the drinking fountain.

That evening two detectives from police headquarters came

to see him, to question him very politely, and he met them smiling.

"Yes, yes," he said. "There was a fat woman in the booth just before me; she left it covered with torn pieces of paper and I brushed them aside into the wastepaper basket. Find out where the basket went, and you'll probably find the money. . . . No, I've no objection at all if you want to look around here just to satisfy yourselves."

Afterward he wondered whether he had not overdone it. They went away apparently convinced, but he couldn't feel safe. He had better leave the money where it was, for a while. There wasn't a chance in a million that anyone would look into the broken drinking fountain. There was no hope of his recovering the money at night: The park gate was locked, and the watch-man was on duty. Someday, when no one was near, he would kneel down as if to tie his shoe . . .

As he entered the park the next morning, he saw something like a red cloud just above the drinking fountain. A red warning of danger. He became very nervous but then saw that it was only a group of toy balloons held by an old man. Lundy had never seen anyone selling balloons here in the three weeks he had lived in the neighborhood. Business couldn't be good for the old man; surely he would leave soon. But when Lundy came that evening, the old man was still in the same spot near the drinking fountain.

Lundy looked at him in passing; he was old, but he looked strong. He might be a younger man in disguise, not a seller of balloons; he might be a detective placed there to watch him. Lundy went home trembling. No one could have seen him put the money away—but suppose that by some accident the money had been found. The police would know the thief must come back for it; thus they had left a man on guard. *But had they left the money there in order to trap him?*

The next morning the balloon-seller was still there. That day Lundy went to the bank and risked a question. No, said the manager of the bank, they hadn't found the money, but they expected to find it. It seemed to Lundy that the manager looked at him in a rather suspicious manner.

That evening he spoke, in passing, to the balloon-seller.

"You work late, eh? Business must be good."

"Not so good. But I stay around until they lock the gates each night and the watchman arrives to guard the place."

There was not a moment when the fountain was not being watched. That was the first night that Lundy could not sleep. In the morning the red cloud was still there, hanging above his treasure.

Well, if business was bad, the balloon-seller would soon leave and go somewhere else to sell his balloons. Lundy waited three more days, in which he saw, morning and evening, that red sign of danger. He couldn't stand this much longer: A balloon-seller, staying in a place where Lundy had never seen one before, couldn't be a balloon-seller. But there was one chance if the police had left the money. Policemen in uniform seldom came here; Lundy could wait for his chance until there was no one around, attack the balloon-seller, knock him out, take the money, and escape before anyone came.

And so he waited for his chance, found the old man alone, walked up to him, pretended to buy a balloon, then hit him straight and hard on the jaw. Down went the old man—down and out; down went Lundy on his knees, his arm reaching into the hole at the base of the fountain.

Up into the air went a dozen red balloons, released from the old man's hand as he fell; a dozen sudden red danger signals which could be seen everywhere in the park and from the nearby streets as well. As Lundy rose, pushing the money into his pocket, he saw a policeman coming up; he turned only to face another, tried to walk away coolly . . .

"Hey!" said the policeman. "What's the matter with old Joe?"

"I don't know. I've done nothing." But the balloon man was talking now, explaining to the policeman what had happened. The policeman turned toward Lundy, severe.

"What's the idea of knocking down an old man that's just out of the hospital?"

"Just out of the hospital?" Lundy asked.

"Sure. He's been sick for a month. Haven't you noticed that for the past month he hasn't been here, at his regular place near the fountain? First time he's been away for twenty years. . . . Here, you—take your hand out of that pocket! Oh, it isn't a gun? Just papers? Well, come along with me and show them to the captain at police headquarters."

1. Why did Lundy go to the bank in the rundown neighborhood?
2. What did he discover when he went into the booth? How did he react?
3. Where did Lundy go after he left the bank?
4. What did he decide to do with the money?
5. What was the "red cloud" above the drinking fountain? Why did this alarm Lundy?
6. What did Lundy suspect that the old man might really be? Why?
7. What plan did Lundy formulate?
8. Why did he go back to the bank? What happened?
9. What happened when Lundy attacked the balloon-seller?
10. What did the policeman think was in Lundy's pocket? What did he order Lundy to do?

*Exercises*

**A.** Use each of the following terms in a sentence:
rundown, investment, a great deal of, stock market, safe-deposit box, to come upon, to overlook, to give up something, to kneel down, to overdo something, to knock somebody out, to put something away, to wait for, to be the matter with.

**B.** Circle the term in parentheses which best completes the sentence.

*Example:* A person who is sick is (well/fat/(ill)/convinced).

1. A man whose duty is to guard a building at night is a (banker/watchman/sergeant/thief).
2. A person who is scared is (relieved/adjusted/balanced/frightened).
3. A (pocket/headquarters/knee/uniform) is part of the body.
4. A thief can also be called a (captain/robber/banker/watchman).
5 To lock something, you need a (key/money/booth/basket).
6. A good place to keep valuable things is a (pillow/hole/drinking fountain/safe-deposit box).

40

7. When you nod, you move your (hand/finger/head/leg) slightly.
8. You use your (finger/elbow/knee/neck) to point.
9. A shop is a (cafeteria / small store / genuine interest / bank).
10. A person with financial trouble has problems with (employees/money/the police/time).

C. A common ending that makes nouns from adjectives is -ity.

> active—activity
> I keep *active* by playing golf. Golf is now my favorite *activity*.

Change the following adjectives to nouns by adding -ity. Then use each word in a sentence.

*Example:* responsibility      She has a job with a lot of *responsibility*.

1. responsible
2. capable
3. similar
4. popular
5. legal

6. creative
7. secure
8. original
9. obese
10. personal

*Discussion*

1. Have you ever been confronted with a great temptation to do something illegal? What was it like?
2. What do you think happened to Lundy after he went to the police station? What would they have charged him with?
3. Do you, or does anyone you know, have a safe-deposit box? Why are safe-deposit boxes used?
4. What does *playing the stock market* involve? What is a *recession?*
5. What does it mean to *act in a suspicious manner?* What mannerisms do people have which might make them appear guilty of something? Do you think you could hide the fact that you had committed a crime?

**41**

# Unit 8: Decision

*Roy Hilligoss*

Even after a year, Tran still called us Aunt Pat and Uncle Andy. We had long since become accustomed to the role of parents, and we assumed that one day we would legally adopt the eight-year-old who was in our care. That was before the letter came from the United Nations Refugee Agency to our home in Fairfax, Virginia, a suburb of Washington, D.C.

Tran had come to us as a refugee from the wars which had been ravaging Southeast Asia for decades. Pat and I had served in the Foreign Service in South Vietnam in the late sixties and then in Kampuchea and Thailand in the seventies and early eighties. We were well acquainted with the suffering of the people of that area—especially of the children.

Tran's parents were alive, but they felt that she would be safer if she went to the United States to live, at least until her homeland became less of a battlefield.

The letter contained news which we had always feared. Tran's parents had been killed in a guerilla raid on their village, which was really an overcrowded camp. Deaths in the camp often came from malnutrition, but this was brutal.

Tran took the news with the slightest show of emotion. We knew that there was a sharp pain in her little heart, but we also knew that she could not show it. Her people were fighting a desperate battle, so she could not let her own individual tragedy supersede the larger one.

"I shall try very hard," she told us in that thin, soft voice, "to be as you would want your own daughter to be. I shall learn American ways as quickly as I can and try to make you proud of me." She managed a weak smile. "I shall even try to admire your history."

We loved Tran and wanted her to stay with us forever. Her sadness was our sadness, and it made us more of a family.

She did well in school. She even reached the finals in a statewide spelling bee. Her accent disappeared, and she became a popular, contributing member of her class in school.

Then a second letter came.

It was from a distant relative of Tran's, one whom she had never met. He was fond of Tran's parents, the letter said, and he would like to offer her a place to live—with his mother. I looked up from the letter. "He says he recognizes the danger of having her return to such a war-torn country, but he thinks Tran would be good company for his mother. She's old and alone. Oh, Pat . . . do you think she'll want to go?"

She shook her head slowly. "I don't know. I've always been afraid that she would leave us one day."

"She doesn't know anyone there anymore. Surely she'd be happier here, with us . . ."

"She's not really ours, Andy. We hoped she would be, and she's tried to pretend, but . . ."

"I know. I was only hoping, not talking sense. Well, let's go up to her room and tell her about this letter."

We watched her glistening brown eyes move quickly back and forth across the page, and when they reached the bottom they stayed there. She was thinking rapidly. Suddenly, I knew she had reached a decision because her face lost all expression—a habit of hers.

"I must go, Aunt Pat, Uncle Andy. It is my duty."

**43**

"It is still dangerous. There are bombs. There is hunger and misery."

"That is why I must go."

"I don't understand," I said.

"I mean . . ." for a moment she paused, ". . . well, when your country's having its most difficult times, that's when it needs you the most."

That sounded a bit too grown-up, too much like something she had read somewhere. I looked at her suspiciously, but her eyes did not waver as they met mine bravely. "All right, Tran. If you feel that's what you must do, then of course you shall go. I'm sorry."

"I'm sorry, too."

We stood there awkwardly for a minute, saying nothing. Then Pat said, "I'm bungry," and we all laughed. It was a mispronunciation of *hungry*. Tran had had a hard time with that word when she first arrived in Fairfax. Now it had become a family word to indicate that we should all go fix dinner together.

I was slicing some tomatoes and cucumbers when I began to miss Tran already, in my heart. How quickly one's world can turn upside down. We didn't talk much about her departure, but we might as well have discussed it constantly. From that first night, on through the week following, there were few signs of cheerfulness in our house.

We decided to drive to New York, where there was a nonstop flight. Pat and I arranged to take off from work, and we worked out the final details of Tran's journey. Her face became less and less revealing of her emotions, but Pat's and mine were a study in tragedy and loss.

On the night before she left, we were sitting in the den, quietly absorbed in our own thoughts. I don't know which of us started it, but soon we were both crying.

"It hurts so much to lose her."

"I know. I know. She's like our own daughter."

The next day came all too quickly, but we knew that we had to leave by noon in order to get to New York's JFK Airport on time for the plane, so we worked diligently to see that all of Tran's bags were packed properly. She insisted on wearing clothes which were similar to those she had worn on the day she arrived. Those clothes had long since been discarded as too small, so we found others in a shop in an Asian section of Arlington. Her face began to betray some emotion as she thrust out her chin defiantly, trying to hide her fears. Then her face became serious, her lower lip pulled slightly in.

"I've been happy here," she said.

We were interrupted by the postal letter carrier, who handed our one piece of mail to Pat. She opened it with a wary look on her face. I noticed that the envelope and handwriting were similar to those of the original letter from Tran's distant relative.

"It doesn't change things," Pat said as she finished the letter. "He says that we won't have to worry about Tran's safety. His mother is being sent to Canada. Well, there won't be any bombs. That's something to be grateful for. Tran is supposed to join his mother in Canada. I guess we'll have to change plans."

Tran, however, was suddenly animated. "Then I don't have to go!"

"You don't? But you said you had to. Don't you want to go now?"

"She was a very old lady, and I thought that she would be alone in the bombings. I thought I would be able to protect her. But now that she's going to be in Canada . . ."

"I never knew that was how you felt," I said in amazement.

She seemed a little embarrassed. "I was afraid. I mean, I thought I could protect her, and I thought my duty was to be with her in a time of trouble. She is part of my family. All these things, but I was afraid, too, and I didn't want to show my fear. Oh, I don't know what I mean, but I know that my decision to stay here is a correct one. Will you let me stay? I want to be a part of this family."

"You're quite a person, Tran," Pat said proudly, beaming through her tears, "and you'll always have a place here in this family. You've made a very good decision."

*Comprehension*

1. Who was Tran? With whom did she live? Where?
2. How had Tran come to live in the United States? Why?
3. What happened to Tran's parents? How did she react to the news?
4. How did Tran adjust to life in the United States?
5. What did the letter from Tran's relative say?

**45**

6. What did Tran decide after reading the letter? What reason did she give for this decision?
7. How did each member of the family behave during the week before Tran's intended departure?
8. Describe the scene just prior to the departure for JFK.
9. What caused Tran to change her mind? What had her true reason for wanting to go away been?
10. What was Tran's final decision? How was it received by Pat and Andy?

*Exercises*

A. Use each of the following terms in a sentence:
long since, to become accustomed to, to be well acquainted with, guerilla raid, malnutrition, to supersede, statewide, spelling bee, to be fond of, war-torn, surely, duty, decision, mispronunciation, to slice, upside down, nonstop, to thrust out, quite a person.

B. A common ending which is used to change adjectives to nouns is *-ness*.

He's always very *kind;* he shows his *kindness* in many ways.

Change the following adjectives to nouns by adding *-ness.* Then use each word in a sentence.

*Example:* seriousness   She always has a look of *seriousness* on her face.

| | |
|---|---|
| 1. serious | 6. polite |
| 2. foolish | 7. eager |
| 3. happy | 8. late |
| 4. sad | 9. quick |
| 5. sharp | 10. close |

C. Match the term in the left column with its OPPOSITE in the right column.

*Example:*      <u>c</u> 7. south      c. north

| | | |
|---|---|---|
| ___ | 1. ashamed | a. late |
| ___ | 2. small | b. sadness |
| ___ | 3. similar | c. north |
| ___ | 4. happiness | d. cowardly |

| | | |
|---|---|---|
| __ 5. brave | e. infrequently |
| __ 6. afraid | f. short |
| __ 7. south | g. large |
| __ 8. often | h. different |
| __ 9. early | i. proud |
| __ 10. long | j. fearless |

*Discussion*

1. Name some countries in Southeast Asia. What languages are spoken there?
2. What is a refugee? Why are there refugees? Where do they go? What are the purposes of refugee camps?
3. What are some difficulties of becoming a citizen of another country? Of adapting to another culture? Of becoming a member of another family?
4. Do you think that Tran's original decision was right? Why?
5. What do you think life is like in Southeast Asia today? In Canada? How do you think life in these areas compares to life in your country? In the United States?

# Unit 9: Ten Steps

*Robert Little*

I put on a clean pullover, a light blue one with a small alligator on the pocket. I got it from the third drawer in the antique chifforobe which we found while on vacation in Maine last summer. I was in our bedroom on the second floor of our suburban house. I could look out the window and see the Cresseys' yard and their dog, Christopher Robin, a large black, friendly mutt. Only the dog wasn't there. I lingered a moment and then went into the small bathroom off the bedroom to wash my face.

The soap was liquid. It came out of a plastic bottle that was called "Soap in a Bottle" or something like that. It was a new product for us; we'd never tried it before. Gina loved to try new gimmicks. I dried my hands and put the hand towel on the left-hand side of the towel rack. I always put it there. A creature of habit, I suppose. The rack looks loose; someday it will fall down and break. The faucet drips because I never got around to replacing a worn washer. I shut the door when I left the bathroom so I wouldn't have to hear the drip-drip-drip.

I went back into our bedroom and looked around the room, taking it all in. On the bed was Gina's stuffed Snoopy dog. She's had that dog for so long now that I forget when and where she got it. I reached down to touch its floppy ears and it fell onto the floor. When I stooped over to pick it up, I noticed a black mark on the wall behind the bed. The bed hits the wall every time someone gets into it, and it makes marks on the wall. I put the dog back onto the bed, but it fell off again as I left the room.

For a moment I couldn't remember the color of the wallpaper in the hallway, so I shut my eyes and tried to remember, tried to picture it. I couldn't. I opened my eyes and saw that, of course, it was that dull green and blue striped design that we had put up when James was still in high school.

I went into James's now-empty room and looked around. We had cleaned out the room completely, so that the only thing I could see was a fly on the windowsill. I opened the window and shooed it out. I then measured the room by walking across from one wall to the other in each direction. It is ten by fourteen.

I went into the family room, with its deep-pile shag rug, and crossed to the desk. Another old-timey piece of furniture. It has a roll-top with dozens of little cubicles and small drawers. Right out in the open, so no one could miss it, I placed a large manila envelope which I had prepared for Gina. I had put instructions, money, deeds, etc., inside. On the outside, I marked "For Gina."

The curtains in this room were red. Where the sun hits them they are now pink, not red. There was a magazine on the coffee table, a recent issue of *Newsweek*. There was also a copy of this week's *TV Guide* on top of the color television. We usually kept it on top of the box which controls our cable TV hookup, but I left it where it was.

The room was functional and comfortable. At one end there were two easy chairs that faced each other. I sat in one, but rose quickly and left the room, heading for the kitchen.

Gina was there looking up a recipe for peach cobbler. She promised Sarah, our neighbor, that she'd make her special cobbler for Sarah's party tomorrow night. She was also defrosting a roast in the microwave for tonight's dinner. I used to like to cook, too, but lately, I've gotten away from it. Gina didn't look up when I entered. She didn't look up as I crossed to leave by the back door, either, but she said, "If you're going to the Giant, will you pick up some whole wheat flour?" I mumbled my assent and went out to the garage. It wasn't connected to the house but was about fifteen yards behind it.

As I walked to the garage, I saw the Risher kid playing with some toys in the sandbox in their backyard. He was running the toy truck back and forth through the sand. The sand was wet and it covered both him and his toys. I watched him, noted his intensity for a few minutes, and then said, "So long, Michael," but he didn't answer. He was too busy with the sand and the truck.

Then I went to the garage and unlocked the door. I walked around our VW Rabbit, running a soft cloth over it, polishing it and cleaning the windows. It had been a faithful car over the years. Not too much power, but steady and reliable. I noticed a few scratches on the door. Then I looked around the rest of the garage at the junk we had accumulated. I made a mental inventory of all the tools, boxes, playthings, patio furniture, and car accessories that seemed to cover every available space in the building. I stood there a few moments longer and then closed the doors and walked up to the front of the house and looked at my watch. It was 9:45.

The front yard had seen a lot of use in its day. At one time, we even had a small vegetable garden there. I think I liked that part of our home the best. It was part of our home, yet it looked out onto Stanford Street and allowed for the rest of the world to pass by. Or to drop over to visit, as many often did.

Around the side of the house we had built some stairs which led down to the side street below. I walked down these steps and counted. Ten. Ten steps. I thought I had counted correctly but wasn't sure. I'd better do it again. Never mind. I walked down the street and looked back at the house. One of the upstairs windows had a shade which was halfway down. I wanted to go back and count the steps again to make sure I had done it correctly, but I didn't.

I walked down to Wisconsin Avenue and then over to Bradley Boulevard and got on the bus. I got off at Montgomery Avenue and walked up to the police station. I asked for Captain Maldonado and told him that if he were still looking for the man who had killed Roger Maguire, then he should arrest me, since I had done it.

Captain Maldonado questioned me for over an hour and then finally said that I should write out a full confession and that if I wanted the assistance of a stenographer, I could have it. He also said that I could make such a confession on a videotape if I so chose. I chose to write it out by myself.

Before I write it all down, about how I killed Maguire and when and where and why, I want to write down the last things I remember about my house and my wife and my neighbors. I'll want to remember them because I won't ever see them again. It's important to me.

*Comprehension*

1. Where are the ten steps that are referred to in the title of the story?
2. What crime has the narrator committed?
3. Why did he carefully observe all the minor details of his house?
4. What details can you remember about his bathroom? Bedroom? Hallway?
5. Who was James? What was in his room? How large is the room?
6. What did the narrator notice in the family room?
7. What was the narrator's wife doing when he went into the kitchen? What did she say to him?
8. Whom did he see in his neighbor's backyard? What was the child doing?
9. What did he do when he went into the garage?
10. How did he feel about the front yard of his house?

*Exercises*

**A.** Use each of the following terms in a sentence:
pullover, antique, mutt, gimmick, towel rack, creature of habit, to picture something, windowsill, to shoo something out, shag rug, roll-top desk, out in the open, cable TV hookup, microwave, so long, reliable, patio furniture, to drop over, confession.

**51**

**B.** A common prefix which is used to change a noun or an adjective into a verb is *-en* (*-em* before *b, m,* or *p*), meaning *make.*

> They want to be *sure* that they are correct.
> They want to *ensure* their correctness.

Change the following nouns or adjectives to verbs by adding *-en* or *-em*. Then use each word in a sentence.

*Example:* enrich    Reading can *enrich* your knowledge.

1. rich
2. able
3. bitter
4. large
5. feeble
6. rage
7. slave
8. power
9. dear
10. act

**C.** Circle the word in parentheses which best completes the sentence.

*Example:* A pullover is a (joke/ (sweater)/hat/pair of socks).

1. A mutt is a (dog/cat/fish/bird).
2. Plastic is a (living/organic/synthetic/breathing) material.
3. We use towels to (eat with/dry ourselves/play with/cover pillows).
4. A faucet is usually found in a (drawer/bedroom/TV/ sink).
5. A room which is 10 by 14 is (24/48/140/280) square feet in area.
6. Directions for preparing a certain type of food or meal are called a (recipe/cobbler/flour/roast).
7. A microwave oven is most likely to be found in a (family room/kitchen/patio/bedroom).
8. The most appropriate place to use patio furniture is (in the living room/in the garage/outside/in a vegetable garden).
9. A fly is a kind of (bird/plane/cloud/insect).
10. Another name for a family room is a (den/living room/ bedroom/garage).

*Discussion*

1. Why did the narrator leave the envelope for his wife? What items, in addition to those mentioned, do you think were inside?
2. What is a family room? What are some of its uses?
3. Describe the various rooms and hallways in your house.
4. Which room in your house do you spend the most time in? Which room does your family congregate in the most? Which is your favorite room?
5. In most of this story, there is nothing but a simple, past-tense description of a man's journey through his home—not much of a plot. When did you first suspect that there was something unusual going on? What made you suspicious?

# Unit 10: The Wrong House

*James N. Young*

The night was dark. And the house was dark. Dark—and silent. The two men ran toward it quietly. They slipped quickly through the dark bushes which surrounded the house. They reached the porch, ran up the steps, and knelt down, breathing heavily, in the dark shadows. They waited, listening.

Silence. Perfect silence. Then—out of the blackness—a whisper: "We can't stay out here. . . . Take this suitcase. . . . Let me try those keys. We've got to get in!"

Ten . . . twenty . . . thirty seconds. With one of the keys, the first man opened the door. Silently, the two men entered the house, closed the door behind them, and locked it.

Whispering, they discussed the situation. They wondered if they had awakened anyone in the house.

"Let's have a look at this place. Careful, Hy. I hope there isn't anybody awake!" And the soft rays of a flashlight swept the room.

It was a large room. A living room. Rugs, carefully rolled, lay piled on one side. The furniture—chairs, tables, couches—was covered by sheets. Dust lay like a light snow over everything.

The man who held the flashlight spoke first. "Well, Blackie," he said, "we're in luck. Looks as if the family's away."

"Yeah, gone for the summer, I guess. We better make sure, though."

Together they searched the house. They went on tiptoe through every room. There could be no doubt about it. The family *was* away. Had been away for weeks.

Yes, Hy Hogan and Blackie Burns were in luck. Only once in the past ten days had their luck failed them. It had been with them when they made their big robbery—their truly magnificent robbery—on the Coast. It had been with them during their thousand-mile trip eastward, by car.

It had been with them every moment—but one.

That moment had come just one hour before. It came when Blackie, driving the car, ran over a policeman. And Blackie, thinking of the suitcase at Hy's feet, had driven away. Swiftly.

There had been a chase, of course. A wild, crazy chase. And when a bullet had punctured the gasoline tank, they had had to abandon the car. But luck or no luck, here they were. Alone, and without a car, in a completely strange town. But safe and sound—with the *suitcase*.

The suitcase lay in the center of the table, in the center of the room. In it, neat little package on neat little package, lay nearly three hundred thousand dollars.

"Listen," said Hogan. "We have to get a car. Quick, too. And we can't steal one: It's too dangerous. We have to buy one. That means that we have to wait until the lots open. That will be about eight o'clock in this town."

"But what are we going to do with that?" Burns pointed to the suitcase.

"Hide it right here. Sure! Why not? It's much safer here than with us—until we get a car."

And so they hid the suitcase. They carried it down to the basement and buried it in an unfinished corner where no cement had been laid. Just before dawn, they slipped out.

As they were walking down the street, Hogan remarked that a Samuel W. Rogers lived in the house they had just left.

"How do you know?"

"Saw the name on some of them library books. The guy's really got a lot of books. Looks like a library in there."

The used car lots opened at eight, as they had supposed. Shortly before nine, Hogan and Burns had a car. A nice little car. Very quiet. Very inconspicuous. Very speedy. They arranged for temporary plates and drove off.

Three blocks from the house, they stopped. Hogan got out. Walked toward the house. He'd just go around to the rear, he thought, and slip in.

Fifty yards from the house, he stopped. Stared, swore softly. The front door was open. The window shades were up. The family had returned!

Well, what bad luck! And what could they do? Break into the cellar that night, and pick up the suitcase? No—too dangerous. Hogan would have to think of something.

"Leave it to me, kid," he told Burns. "You drive the car. I'll do the special brain work. Let's find a telephone. Quick!"

Ten minutes later, Hogan was consulting a telephone directory. Yes, there it was—Samuel W. Rogers, 555-6329.

A moment later he was talking to the surprised Mr. Rogers.

"Hello," he began, "is this Mr. Rogers—Mr. Samuel Rogers?"

"Yes, this is Mr. Rogers."

Hogan cleared his throat. "Mr. Rogers," he said—and his tone was sharp, official, impressive—"this is Headquarters, Police Headquarters, talking. I am Simpson. Sergeant Simpson, of the detective division . . ."

"Yes, yes!" came over the wire.

"The Chief—the Chief of Police, you know," here Hogan lowered his voice a little—"has ordered me to get in touch with you. He's sending me out with one of our men to see you."

"Am I in trouble of some kind?" asked Mr. Rogers.

"No, no, no. Nothing like that. But I have something of great importance to talk to you about."

**56**

"Very well," came the voice of Mr. Rogers. "I'll wait for you."

"And, Mr. Rogers," Hogan cautioned, "please keep quiet about this. Don't say anything to anybody. You'll understand why when I see you."

On the way back to the house, Hogan explained his idea to Burns.

Within ten minutes, "Sergeant Simpson" and "Detective Johnson" were conversing with the surprised Mr. Rogers. Mr. Rogers was a small man. Rather insignificant. He had pale blue eyes. Not much of a chin. A funny little face. He was nervous —a badly frightened man.

Hogan told the whole story. Somewhat changed, of course. Mr. Rogers was surprised, but he was delighted to be able to help the police.

He accompanied Hy Hogan to the cellar. And together they dug up the suitcase. Took it to the living room, opened it, saw that it had not been touched—that it really did hold a small fortune. Bills, bills, bills!

Hogan closed the suitcase.

"And now, Mr. Rogers," he announced, in his best official manner, "Johnson and I must run along. The Chief wants a report—quick. We have to catch the rest of the robbers. I'll keep in touch with you."

He picked up the suitcase and rose. Burns also rose. Mr. Rogers also rose. The trio walked to the door. Mr. Rogers opened it. "Come on in, boys," he said pleasantly—and in walked three men. Large men. Strong men. Men in police uniform who, without fear, stared at Hy Hogan and Blackie Burns.

"What does this mean, Mr. Rogers?" asked Hogan.

"It's quite simple," said Mr. Rogers. "It just happens that *I* am the Chief of Police!"

*Comprehension*

1. In the first few paragraphs of the story, how do we know that the two men are breaking into the house?
2. What did they conclude about the residents of the house? Why?
3. What caused them to abandon their car?
4. What was in the suitcase? Where and how did Hogan and Burns get it?

**57**

5. Where did they decide to hide the suitcase?
6. When did they buy a car? Where? What was it like?
7. What did they discover when they returned to the house?
8. Explain Hogan's plan to get the money back.
9. What did Hogan and Burns do when they got the suitcase back upstairs?
10. Why didn't their plan succeed?

*Exercises*

**A.** Use each of the following terms in a sentence:
to kneel down, to be in luck, to make sure, no doubt about it, to be away, to run over, safe and sound, dawn, to slip out, inconspicuous, to get in touch with, to be in trouble, to keep quiet, insignificant, trio, on the way back.

**B.** Some nouns ending in *-tion* or *-ment* may be changed to adjectives by adding *-al*.

> She went abroad for her *education*.
> She had some *educational* experiences while she was abroad.

Change the following nouns to adjectives by adding *-al*. Then use each word in a sentence.

*Example:* environmental    Air pollution is an *environmental* problem.

1. environment
2. government
3. vocation
4. division
5. recreation
6. tradition
7. ornament
8. occupation
9. conversation
10. addition

**C.** Match the term in the left column with the term which has a SIMILAR meaning in the right column.

*Example:*    _a_ 4. steps    a. stairs

___ 1. robber          a. stairs
___ 2. insignificant   b. fast
___ 3. abandon         c. make a hole
___ 4. steps           d. desert
___ 5. terrified       e. middle
___ 6. puncture        f. speak softly

58

|  |  |  |
|---|---|---|
| — | **7.** silly | **g.** thief |
| — | **8.** center | **h.** unimportant |
| — | **9.** speedy | **i.** foolish |
| — | **10.** whisper | **j.** frightened |

*Discussion*

1. Many of the sentences in this story aren't true sentences; for example, in the second paragraph: "Silence. Perfect silence." What quality does this writing technique add to the story?
2. What can you tell about the character and education level of Hogan and Burns from their dialogue?
3. What else might the robbers have done to retrieve the suitcase? What would you have done?
4. What would you have done if you had been Chief of Police Rogers?
5. What precautions do you or your family take at home when you go away on vacation?

# Unit 11: Irene's Sister

*Vina Delmar*

It had been fifteen years since I'd seen my friends on Rigel Space Colony IV. I looked forward to my return. This is a story of those days, fifteen years ago.

It was the year 2089, the year the schools didn't open on time, the year the plague of space fever descended and caught us. We were as defenseless as if we were inhabitants of some medieval city faced with a new and terrible sickness.

I was only a young girl at the time. My friends and I were confused and frightened, but our parents had no answers. They were as terrified as we. "All we know is that it's some kind of space fever," they told us. "It either kills you or leaves you crippled forever. Don't get too close to anyone or anything strange. You can't be too careful."

Fear held us so completely that we forgot how to laugh or

**60**

play. I can remember lying in bed at night waiting for the disease to strike me. I had no idea what form it might take and I lay very quietly, praying that when I next wished to move my legs or arms, I would be able to do so as I had always done in the past.

There was one among us, however, who had no fear of the terrible plague. That girl was Irene Crane. In my mind's eye, I can still see her as she was back there in those difficult days. She was a yellow-haired child with a happy ring to her laughter and the greatest capacity for fun of anyone I've ever known. She was the school beauty, popular with teachers and students alike; and if she was not the most intelligent of our group, that was easily forgiven, for one does not expect to find genius in a flower.

Irene had a sister who was a year younger. Her mother called her Caroline, but outside the house she was known simply as Irene's sister. It was natural for her to be Irene's sister, just as it was natural for us to be a nameless group of girls known as Irene's friends. Irene was the center of our small world, and we revolved about her brilliance and asked for no recognition for ourselves. Irene's sister, conscious of her inability to compete with the beauty and entrancing manner of Irene, was perfectly content to be only a pale reflection of our yellow-haired commander.

Only once were we at odds with Irene's way of thinking. That was when she said, "I'm not scared of that space fever. There's no way any of us will ever get it. You'll see. None of us will. We're invincible."

None of us agreed with her. We were a little ashamed of our fears of the terrible sickness which had befallen our small planet, but those fears were there just the same. They were with us day and night.

At least we had each other. Since we played together, went to school together, and knew each other since early childhood, there was nothing to fear by spending time together.

I can remember the day that we all went over to Ginny Smith's house for games and light refreshments. For our health's sake, the grown-ups looked upon the party with some doubts, but for the good of our morale they consented.

"After all," they said to one another, "it's the same group of girls who see each other almost every day anyway. It'll be all right."

"It's the same group except for Irene's sister." She hadn't been invited because she was not in our grade at school, and Ginny Smith hadn't known that Irene had a sister.

"It doesn't matter," Irene said. "Caroline isn't feeling well. She has an upset stomach, I guess."

The games were fun. We played three-dimensional electronic chess, and there was a hilarious round of hide-behind-the-time-warp which left all of us laughing. The food was great, too. Ginny's parents are really wonderful cooks. We also danced. Of course, Irene was the group's best dancer. She taught us a new rock dance which she had learned on her recent trip back to Earth with her grandparents. We listened to that wonderful singing group, Cherie and the Asteroids.

It was a beautiful day. We all seemed to forget for a while that something strange and terrible walked everywhere on our small colony deep in space. We forgot for a few hours the dangers that lay beyond Ginny's dome house. We were just getting ready to leave and thanking Ginny for a lovely party when the video-phone beeped.

I can still see Ginny's mother as she stood talking to Irene's mother on that phone. There was a look of horror on both women's faces. I can still see the tears in their eyes. We couldn't hear what they were saying, probably because we wanted to avoid and ignore what was obviously bad news.

"Irene," she said in a choked voice, "that was your mother. Your sister has caught the space fever. You can't go home; it's too dangerous. You'll have to stay here." There was a horrible pause. Then, "It's too late for us to be afraid of you, child. You've been here all day."

We went away without touching Irene, some of us without speaking to her. The plague had reached out and struck at us. We hurried home afraid of each other, ashamed of our fear, and unable to keep back the thought that tomorrow we would all be attacked by death or lameness.

Irene stayed with the Smiths, I suppose. I don't know. I hurried home and wrote an emotional, crazy little letter. My father had recently been transferred to a more distant space colony. He was an engineer with the United Earth Command. He specialized in gravity problems and in solving them. He and my mother had gone off to the new planet and left me with an aunt and uncle so that I could finish school on Rigel. Now I no longer wanted to stay. My letter begged my father to come and get me and take me to safety somewhere, anywhere. I did not know that the plague was widespread. I thought it was only on our colony. Anyway, both my parents came and took me away. I went happily and thankfully, not knowing that it would be fifteen years before I set foot on Rigel again.

On my first night back, which was last night, I stayed with an old friend, Maureen Morales. I was surprised to find that Maureen's living room was set up for a party.

"Just the old group," she explained, "and their husbands and boyfriends. You remember Ginny Smith; she'll be here. So will Lila Day, Sally Mullen, the Crane sisters, and the rest of the old group."

A strange feeling of terror ran through me at the mention of the Crane girls. I was a child again, frightened before a terrible, mysterious force that wanted to kill me.

"I remember them all," I said. "How are the Crane girls?"

"The same as ever, just exactly the same. One popular and one a complete failure."

"That's not fair," I protested. "Caroline had the space fever. I'm surprised that she's even alive. How can you expect her to . . ."

"But it's Irene who's the failure. She's even a little ridiculous for a grown woman. Remember how she used to laugh and play jokes and be the life of any party? Well, she's still the same, only the jokes are old and stale, and everything she says comes out sounding silly. But we have discovered that you can't invite Caroline without inviting Irene, so . . ."

"But, you mean that Caroline is well?"

"Of course she is. She had good care and good sense used on her, and she's as fine as anyone. A lot finer, I guess. She went through so much pain and suffering that she has more depth and understanding than most people. She's so strong and dependable. Of course, she thanks her doctor and her nurse and her mother for everything, and they say that it was Caroline's patience and courage that helped them to help her. Wait till you see her. She's—"

It was at that moment that the front door scanner told us that someone was approaching. Maureen's husband saw it first, but

**63**

he was busy preparing a punch for the party, so he called out, "Maureen, would you get the door, please? It's Caroline's sister."

*Comprehension*

1. Where does this story take place? When?
2. What is the "plague" that the author refers to in this story?
3. Why was Irene Crane so popular as a child?
4. How did Caroline feel about her sister?
5. What did the girls do at the party at Ginny Smith's house?
6. Why wasn't Caroline at the party?
7. What did Mrs. Crane call Mrs. Smith about?
8. Why did the narrator write to her father? What was the result?
9. What had Maureen arranged for the narrator's return to Rigel? Who was going to be there?
10. What is the significance of the last sentence in the story?

*Exercises*

**A.** Use each of the following terms in a sentence:
to look forward to, on time, medieval, to be at odds with, genius, to be ashamed of, for the good of, grown-up, after all, upset stomach, three-dimensional, colony, dome, gravity, widespread, to set foot on, to be set up for, the life of the party.

**B.** Many English words can be used as either a verb or a noun without any change in form.

> We're going to *record* our voices on that tape.
> Did you play the *record* of that new rock group?

Use each of the following words first as a verb and then as a noun.

| | |
|---|---|
| 1. work | 6. lock |
| 2. help | 7. light |
| 3. experiment | 8. hurry |
| 4. play | 9. dream |
| 5. watch | 10. knock |

**C.** Fill in the blanks with a word from the list.

> *Example:* All of my parents' children are girls; I don't have any *brothers*.

| aunt | grandfather | nephew |
|------|-------------|--------|
| brother | grandmother | sister |
| brother-in-law | great-uncle | sister-in-law |
| cousin | niece | uncle |

1. My sister's daughter is my _____.
2. My father's mother is my _____.
3. Your mother's brother is your _____.
4. Your grandfather's brother is your _____.
5. If your mother's sister had a child, that child would be your _____.
6. My father's sister is my _____.
7. I hope our baby is a girl; then our son will have a _____.
8. When I got married, my husband's sister became my _____.
9. My brother's son is my _____.
10. Your mother's father is your _____.

*Discussion*

1. Do you think people will be living on other planets or in space colonies in the next century? Why?
2. What diseases are as dangerous and as frightening as the one in the story?
3. Why is it that adversity, such as a severe illness, seems to make some people stronger?
4. Do you know any brothers or sisters who are like Irene and Caroline (when they were young)? What is their relationship like?
5. The story mentioned some household products (videophone, front door scanner, etc.) which might be used in the future. What other developments do you envision for our future?

# Unit 12: Detour to Romance

*Gilbert Wright*

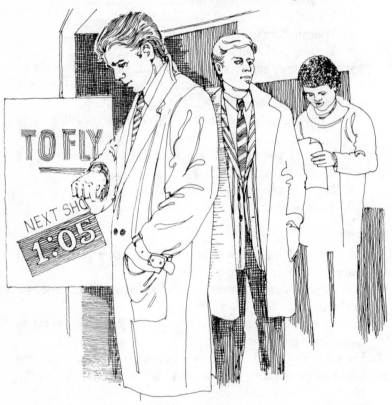

The Air and Space Museum, which is part of the Smithsonian Institute in Washington, D.C., is the most visited museum in the United States. Year after year, more people visit this massive exhibit honoring the men and women who have pioneered flight and the exploration of space than visit any other monument or museum in the entire country.

I work in a little room off the main entrance to the museum, checking coats and other articles which people do not want to carry around as they tour the building. I see virtually everyone who enters the museum.

Maryanne Wilson, who used to sell souvenirs at the stand

located next to my checkroom, has studied the laws of probability because she likes to bet. She claims that she can calculate, according to her system, the odds against anything happening. She calculated once that if I held my job for 112 more years, I would know everyone in the United States by sight.

I myself came to the conclusion that at the very least, if I waited long enough, I would see everyone who traveled. I've told people this theory for years, but no one ever did anything about it. No one except Sidney, that is. Sidney came into the museum a little over three years ago.

There are several short films which are shown at the museum every half hour. The one called *To Fly* is the most popular, and people line up hours in advance to see a particular showing. Sidney was waiting at the head of the line for the 1:05 showing one afternoon. He was standing there looking very nervous.

I remember noticing him that first day. He wasn't much more than a thin, anxious kid, but there was something about him. It was eerie. I knew. I just knew that he was meeting his girlfriend and that they were going to go off and be married that same day. There's no use in my trying to explain how I knew this, but after one has watched people for eighteen years, as I have, it's easy.

Well, more tourists poured through the front door, so I got busy. I didn't look up again until it was nearly time for the 1:35 showing of *To Fly*. I was surprised to see that the young fellow was still there, at the head of the line.

Sidney's girlfriend wasn't there for the 2:05 either, nor the 2:35, and when the viewers of the 3:05 showing were leaving the theater, Sidney was looking pretty desperate. Soon he wandered over near my window, so I called out and asked him if I could help.

He described her in a loving way. "She's small and dark, nineteen years old, and has a spirited face. I mean she can get mad, but she never stays mad for long. She has a short coat made of soft brown leather, but maybe she's not wearing it."

I couldn't remember seeing anyone like her.

He showed me a letter, actually a postcard, from her: "I'll be there Thursday. Meet me at the museum. Let's fly. Love, love, love, love, Kate." It was from Omaha, Nebraska.

"Why don't you phone home? She's probably called there, since she missed you here."

He looked ill. "I've only been in town two days. We were going to meet and then drive to Florida, where I've got a job promised me. I have no address." He touched the postcard. "I got this general delivery." And with that, he walked back to the head of the line to look over the people going to the 4:35 show.

When I came on duty the next day, he was still there.

"Did she work anywhere?" I asked.

He nodded. "She was a gardener. I called her former boss, but all he knows is that she left to get married."

Well, that's how it began. Sidney hung around that line and the museum for the next three or four days. The D.C. police looked into the case, but they couldn't do much; after all, no crime had been committed. Maybe she had just changed her mind, they reasoned. Somehow I didn't believe that.

One day, after about two weeks, I told Sidney of my theory. "If she's a traveler, and if you wait long enough, you'll see her coming through that door someday." He turned and looked at the front entrance as though he had never seen it before, while I went on explaining about Maryanne's figures on the laws of probability.

Sidney went to work for Maryanne as a clerk. "I had to get a job somewhere, didn't I?" he said sheepishly. Neither of us ever spoke of Kate anymore, and we dropped the subject of the laws of probability, but I noticed that Sidney observed every person who entered that most visited of all museums.

Maryanne tired of life in the nation's capital about a year later, and she moved to New Mexico. Sidney took over the stand, expanded it, and soon had a very nice little business.

Then came yesterday. It was spring and the tourists had descended on Washington, as they do every year. There was an endless stream of them, as usual. What made yesterday different began with a great noise.

Sidney cried out and the next thing I knew, there were souvenirs and cards, dolls, and who-knows-what flying all over the place. Sidney had leaped over the counter and upset everything in sight. He ran across the floor and grabbed a young woman who was standing not ten feet from my window. She was small and dark and had an interesting face.

For a while they just hung on to each other, laughing and

crying and saying things which had no meaning. She'd say a few words like, "It was the other one, the one down the street, the one called the castle on the mall," and he'd kiss her speechless and tell her the many things he'd done to try to find her. What apparently had happened, three years ago, was that Kate had gone to a different building. When she was young, her family had taken her to a part of the Smithsonian where the plane in which Charles Lindbergh had flown across the Atlantic Ocean was. She remembered where it was, but she didn't think that they would move it. She had waited at another museum for days and had spent all her money trying to find Sidney. Finally, she got a job as a gardener with the Department of the Interior, working on the grounds of various government buildings around town.

"You mean you've been here all the time?" Sidney gasped incredulously.

She nodded.

"But everybody visits the Air and Space Museum! You mean you've been here for three years and never come through those doors before? I've been here all the time, waiting and waiting for the day to come, watching everyone who came here. . . ."

She began to look pale. She looked over at the doors and said in a weak voice, "No, I've never been in here before. But Sidney, for almost three years I've been working on the grounds around this very building! I've thought about coming in here often, but I never did before today. I just never got around to it." Then she threw her arms around him and they both began to cry again.

What a wonderful drama had unfolded before my eyes! It's too bad Maryanne couldn't have seen it, too. The wonderful thing is how the laws of probability worked so hard and so long until they finally got Kate to walk through those front doors of ours.

*Comprehension*

1. Who was Maryanne? Why did she know about the laws of probability?
2. Why was Sidney at the Air and Space Museum?
3. Where is the museum? What is in it? Do many people visit it each year?

4. What is *To Fly?* How often is it shown? Why is there often a line?
5. Why were Sidney and Kate unable to get in touch with each other?
6. Were the police able to help Sidney? What did they reason? What did the narrator of the story believe?
7. Where did Sidney get a job? Why?
8. Where did Maryanne go? Why?
9. What happened when Sidney saw Kate for the first time in three years?
10. Why hadn't they met three years ago? What had Kate been doing for the past three years?

*Exercises*

**A.** Use each of the following terms in a sentence:
year after year, to pioneer, monument, to carry around, to know by sight, at the very least, every half hour, eerie, spirited, general delivery, to look over, to hang around, sheepishly, to drop the subject, endless stream, incredulously, grounds, to unfold.

**B.** Circle the term in parentheses which best completes the sentence.

*Example:* A small token of remembrance can be called a (movie/museum/souvenir/castle).

1. If you grab something, you (install it/wait for it/go up to it/seize it roughly).
2. The best place to leave packages for a while is in (a garage/a magazine stand/a checkroom/an elevator).
3. A person who is mad is (small and dark/neat/excited/angry).
4. We use (lamps/couches/sheets/dreams) on a bed.
5. Someone walking on tiptoe is probably trying to walk (well/quickly/carelessly/quietly).
6. Grown-ups are (children/youngsters/adults/parents).
7. An enchanting person is (healthy/scared/frightened/charming).
8. If you whisper, you are (speaking softly/speaking loudly/yelling/crying).

9. Someone who does something silently does it (impressively/well/quickly/noiselessly).
10. He wanted to borrow some money from me, but I didn't want to (ask/lend/need/beg) him any.

C. Adjectives of three or more syllables form their superlatives by the addition of *most*.

> sentimental    He is the *most sentimental* man I have ever met.

Use *most* to form the superlative of each of these adjectives. Then use the superlative form in a sentence.

1. beautiful
2. typical
3. natural
4. exciting
5. popular
6. professional
7. official
8. wonderful
9. successful
10. horrible

*Discussion*

1. Describe a time when you missed someone because you got your directions or instructions wrong. How did you find the person?
2. What are the laws of probability? How can they be helpful to people?
3. Would you enjoy visiting the Air and Space Museum? What other places of interest do you know about in Washington, D.C.?
4. What kinds of souvenirs do you buy when you visit an interesting place?
5. What does a gardener do? Do you enjoy gardening?

# Unit 13: Final Break

## Ian S. Thompson

Carson and Jane had been living together for eleven years. Alone. Just the two of them. He had been there whenever she needed him, taking care of her when she was sick, giving her emotional support, encouraging her in whatever she wanted to do.

Now she was going to leave.

They had been walking along Kingsford Street, not touching. Jane stopped and put her hand on his arm.

"This is the new bookstore I was telling you about. I thought you might find the sort of thing we spoke of here. I want to get something for you."

Carson nodded as though in a daze. There were tears in his eyes as he looked into the bookstore window. The purchase had been her idea, not his. He had wanted to explore this new discount store, but these circumstances were hardly conducive to browsing.

"What about this new Robert Ludlum thriller?" Jane pointed to a display near the front counter. "You've always enjoyed his work."

Carson trembled a little. One of the little things he loved so much about Jane was the truly genuine interest she had always taken in what he liked to read. It had made him feel young and somehow, loved, though in his heart he knew he was no longer young.

"Yes, yes, I have always enjoyed him, haven't I? He seems to come out with a new novel just frequently enough. I've always said that, haven't I?" He was feeling a bit self-conscious, and he avoided meeting her eyes because there was so much in his own that he didn't want her to see.

They continued looking up and down the aisles, bantering about the latest exposés by Hollywood stars, speculating on certain politicians' chances in upcoming elections as a result of their books, picking up whatever book caught their eyes.

By the time Jane began reading cartoons from the latest Garfield book, Carson was wishing that they had never come into the store in the first place. But Jane had been insistent. She wanted to give him something. A parting gift, she had called it.

She was smiling at him across a case of books, smiling out of green, clear, untroubled eyes. It surprised him. And yet, why should it, he asked himself as he aimlessly turned the pages of the new cookbook he was holding. He wasn't looking at anything on the pages. He had always tried to be modern, hadn't he, and modernity (or at least part of it) was to see these things through bravely, when and if they came.

These have been happy years, he thought. Oh, there were some ups and downs, but through it all, he'd been truly happy.

Five minutes later, they were out on the street again, out in the sunshine. After looking at her watch, Jane suggested an afternoon cocktail. There was an expression of excitement in her eyes which Carson couldn't quite place. "I know a place," she said, "You'll like it there."

It was a small, ordinary pub on one of the side streets. She ordered the house special for both of them.

She didn't speak, but she leaned across the table and took his hand. She smiled pleasantly and sincerely.

He didn't know what to make of it at first. Then he found himself striving to keep his wits about him.

They drank their cocktails quickly, then sat back as though relaxed, and looked off into space.

"You really should reconsider staying in the house alone. It's been too big for the two of us; surely you'll get lost rattling around there all by yourself. I feel bad about this. I wish there were something I could do to make it easier."

There was one thing he could have suggested, but his pride wouldn't let him. Besides, he didn't want her to have any feelings of regret, any pangs of conscience. It would taint all those wonderful years together.

"No," he said. "I'll be fine. Really. I'm the original Mr. Tough Guy, remember?"

But she didn't want to let the subject drop. She changed directions. "Another thing. I've been reluctant to mention this before now, but I want you to know that you'll never have to worry about money, if that ever becomes a problem. I know you're sensitive about that area, but I feel strongly about it, and I want you to know . . ."

"OK, OK. I understand what you're saying. And you're right. I am sensitive about money. I've always paid my own way. It's true, I have some difficulties now, but I don't expect you to do anything."

"Why not?" She brushed his arguments aside, "And John (she mentioned his name for the first time) agrees. We were talking about it last night."

*John . . . we . . .* How easily, how familiarly she said those words. And yet two months ago, they hadn't even met. Two months . . . was it really only two months since she'd gone on that job interview to Denver? A lot can change in a short time.

He'd realized after she'd returned that something had changed. It wasn't anything tangible, it was more of a feeling than anything else. They hadn't spoken of it, but he'd had a deep instinct that she was no longer his, that he was sharing her with someone else. Someone special to her.

He tried to imagine this John. Young, no doubt. Young, virile, and vibrant. Probably a good dancer or musician; she'd be attracted to qualities like those. And there was the appeal of the city, too. Life on the ranch was a good life, but he realized that it could also be dull for an attractive young woman. His thoughts drifted back to John. He hadn't met him yet, even though he had known of his existence for two months now.

John . . . He worked in the insurance business, she had told him. And smart, too; a clever, intelligent young man. Not that it mattered to Carson. Love was the key. If he loved her as deeply as Carson did—that was all that mattered. He was glad that the young man was smart and clever, but he was mostly concerned with her emotional needs.

Would he work to keep her happy? Would he be considerate of her, kind to her? Would he look at her with honest eyes like those of that young man over there? His attention was drawn to a handsome man who had just entered the restaurant. The fellow was looking around, hesitantly.

Then the young man stopped and looked directly at their table. Carson returned his gaze, his eyes widening in surprise as Jane rose to her feet. The lad was moving toward their table.

"So, you were able to get here after all, darling!" He heard Jane's voice greet him. They embraced briefly and then she turned to Carson. "Dad, this is John! This is the lucky bridegroom who's going to sweep your daughter off to the altar next week!"

*Comprehension*

1. What is the significance of the title of this story?
2. What was the relationship between Jane and Carson like?
3. What were they doing in the bookstore?
4. What kinds of books did they look at?
5. Where did they go after they left the bookstore? Why?
6. What was Jane's concern regarding Carson's living arrangements?
7. What had happened in Denver? How did Carson feel about it?
8. What did John do for a living? What kind of person was he?
9. What happened in the pub when John walked in?
10. Before the last paragraph, was there some doubt in your mind as to the relationships among the three characters? Why?

**A.** Use each of the following terms in a sentence:
emotional, to walk along, in a daze, to be conducive to, in one's heart, self-conscious, exposé, upcoming, to catch one's eye, in the first place, when and if, house special, to keep one's wits about one, to let a subject drop, to pay one's own way, considerate.

**B.** The ending *-ive* can change some verbs into adjectives.

They *act* in a very busy manner. They're *active* people.

Change these verbs into adjectives by adding *-ive*. Then use each word in a sentence.

*Example:* impressive      His experience was *impressive* enough to get him the job.

1. impress
2. construct
3. prevent
4. create
5. repress
6. express
7. correct
8. instruct
9. object
10. attract

**C.** Direct discourse quotes the exact words spoken. Indirect discourse quotes words indirectly.

*Direct:*      Jane said, "I know a good place."
*Indirect:*    Jane said she knew a good place.

Change these sentences from direct to indirect discourse.

1. She said, "You'll like it there."
2. He said, "You've always enjoyed your work."
3. I said, "I want to get something for you."
4. I said, "You shouldn't get anything for me."
5. You said, "I want to meet John."
6. You said, "He agrees with me."
7. They asked us, "Do you want to buy these books?"
8. Jane said, "I don't want you to live here alone."
9. She said, "Don't you want to talk about this?"
10. We told her, "You're very sensitive about this."

*Discussion*

1. Do you think fathers are often upset when their daughters decide to marry? Why?

2. What would this story have been like if Carson were a mother and Jane were her son?
3. Where and how will (did) you introduce your fiancé (fiancée) to your parents?
4. Are you close to your opposite-sex parent? Will it be (Was it) difficult to leave that parent in order to get married?
5. What are *cocktails*? How many can you name?

# Unit 14: A Case of Suspicion

*Ed Wallace*

He threw back the covers and sat up on his bed, his feet feeling along the cold floor for his house slippers, the telephone ringing insistently a little distance away.

He turned on the light and picked up the phone.

"This is Dr. Benson," he said.

The November wind was bringing sounds of winter as it blew around the little white house. The doctor got into his clothes, went to the table, and stared a moment at his watch, his spirit complaining at the job ahead of him.

2 A.M.

His mind complained at the hour and at why people in such remote, rural parts of the country chose such improper times to be born. He picked up his two satchels: the short pill bag, as the people of the town knew it, and the long obstetrical case—the baby bag, they called it.

He debated whether to bring his cigarettes. He knew he should stop, knew he was setting a bad example for people—a doctor smoking! Imagine! But old habits die hard. He put the pack in his pocket. The cold wind felt like a surgeon's knife at his face as he opened the door and ran, bending low, around the driveway to the garage.

His car started with difficulty, coughed a half-dozen times as he drove down the driveway, but then began to run more smoothly as he turned down Grass Street and onto the deserted highway.

Mrs. Ott Sorley, whom Dr. Benson was on his way to visit, already had almost a dozen children, but it seemed to the doctor that never once had she had a baby in good weather, nor in daylight. And while Dr. Benson was a country doctor, he was still a young man and couldn't find the pleasure that his father, "the old Doc Benson," had found in seeing Ott, the father, always two or three babies behind in the payment of his baby bills.

It was a long ride out to the Sorley farm and the sight of a man walking alone along the country road, as seen just ahead by the lights of the car, was a welcome relief to the doctor. He slowed down and looked at the man walking along with difficulty against the wind, a little package under his arm.

Coming alongside, Dr. Benson stopped and invited the man to ride. The man got in.

"Are you going far?" asked the doctor.

"I'm going all the way to Detroit," said the man, a rather thin man with small black eyes filled with tears from the wind. "Could you give me a cigarette?"

Dr. Benson unbuttoned his coat, then remembered the cigarettes in the outer pocket of his overcoat. He took out the package and gave it to the rider, who then looked in his own pockets for a match. When the cigarette was lighted, the man held the package a moment, then asked, "Do you mind, mister, if I take another cigarette for later?" The rider shook the package to remove another cigarette without waiting for the doctor to answer. Dr. Benson felt a hand touch his pocket.

"I'll put them back in your pocket," the little fellow said.

Dr. Benson put his hand down quickly to receive the cigarettes and was a little irritated to find them already in his pocket.

After a few minutes, Dr. Benson said, "So you're going to Detroit?"

"I'm going out to look for work in one of the automobile plants."

"Are you a mechanic?" asked the doctor.

"More or less. I've been driving a truck since the Vietnam War ended. But I lost my job about a month ago."

"Were you in the army during the war?"

"Yeah, I was a medic. Used to fly in medical helicopters right up into battle conditions. I saw a lot of action."

"Is that so?" said Dr. Benson. "I'm a doctor myself. Benson is my name."

"I thought this car smelled like pills," the man laughed. Then he added, more seriously, "My name is Corrigan."

They rode along silently for a few minutes, and the rider moved himself in his seat and placed his package on the floor. As the man leaned over, Dr. Benson caught his first good look at the small, catlike face.

The doctor also noticed the long, deep scar on the man's cheek, bright and red-looking as though it were of recent origin. He thought of Mrs. Ott Sorley and reached for his watch. His fingers went deep into his pocket before he realized that his watch was not there.

Dr. Benson moved his hand very slowly and carefully below the seat until he felt the leather holster in which he had an automatic handgun. Ever since he was robbed at gunpoint eighteen months ago, he always traveled with his gun under the seat. Especially late at night.

He drew the pistol out slowly and held it in the darkness at his side. When he had to slow down for a sharp bend in the road, he stepped on the brake hard and pushed the nose of the gun into his rider's side.

"Put that watch into my pocket," he said angrily.

The rider jumped with fear and put up his hands quickly. "My God, mister," he whispered. "I thought you . . ."

Dr. Benson pushed the pistol still deeper into the man's side and repeated coldly, "Put that watch in my pocket before I let this gun go off."

Corrigan reached for it, and with trembling hands, tried to put the watch into the doctor's pocket. With his free hand, Dr. Benson pushed the watch down into his pocket. He opened the door and forced the man out of the car.

"I'm out here tonight, probably to save a woman's life, but I took the time to try to help you," he said to the man angrily.

Dr. Benson started the car quickly and the wind closed the door with a loud noise. He put the pistol back into the leather holster under the seat and hurried on.

The drive up the mountain to the Sorley farm was less difficult than he had feared, and Ott Sorley had sent one of his older boys down the road with a lantern to help him across the old wooden bridge that led up to the little farmhouse.

Mrs. Sorley's many previous experiences with bringing children into the world apparently helped her greatly because she delivered this child with little difficulty, and there was no need on Dr. Benson's part for the instruments in the long bag.

When it was over, Dr. Benson gave in to his vice and sat down for a cigarette.

"A fellow I picked up hitchhiking on my way here tonight tried to rob me," he said to Ott. "He took my watch, but I somehow summoned my courage and pulled my gun on him. He quickly decided to give it back to me."

Ott smiled wide at such an exciting story coming from young Dr. Benson.

"Well, I'm glad he gave it back to you," Ott said. "Because if he hadn't, we wouldn't have any idea what time the child was born. What time would you say it happened, Doc?"

Dr. Benson took the watch from his pocket.

"The baby was delivered about thirty minutes ago, and right now it's . . ." He walked over to the lamp on the table.

He stared strangely at the watch in his hand. The crystal was cracked and the top was broken. He turned the watch over and held it closer to the lamp. He studied the worn inscription:

"To Corporal Tim Corrigan, Medevac Unit, whose personal bravery preserved our lives the night of Nov. 3, 1971, near Saigon on the Mekong River. Nurses Hohorst, Walsh, and Bryan."

*Comprehension*

1. At what time was Dr. Benson awakened by the phone? What time of year was it?
2. Who had called him? Why? How did he feel about it?
3. How did he feel about his smoking habit?
4. Why did Dr. Benson stop his car on the way to the Sorley farm?
5. What did the rider do with the doctor's pack of cigarettes?
6. What did the man do in the army?
7. How did the doctor get the watch from the man?
8. Was the delivery of the Sorley child difficult? Why?
9. Why did the doctor take the watch from his pocket?
10. Explain the irony of the story, particularly after the doctor read the inscription on the watch.

*Exercises*

**A.** Use each of the following terms in a sentence:
to sit up, to turn on, bad example, old habit, to slow down, to mind, more or less, helicopter, to see action, catlike, at gunpoint, automatic, bend in the road, to bring children into the world, to give in to, to summon one's courage.

**B.** A common ending among adverbs is *-ly*.

We always take *careful* notes in class.
We always take notes in class *carefully*.

Change the following adjectives to adverbs by adding *-ly*. Then use each word in a sentence.

*Example:* constant     The baby cries *constantly*.

1. constant
2. eager
3. slow
4. improper
5. silent
6. serious
7. profound
8. identical
9. fortunate
10. intimate

**C.** Circle the word in parentheses which best completes each sentence.

*Example:* A holster is a case used to hold (fruit/medicine/a watch/(a gun)).

1. If a telephone rings insistently, it rings (off and on/ seldom/now and then/steadily).
2. A receiver is part of a (telephone/medicine bag/car/pistol).
3. My car needed to be repaired, so I took it to a (farmer/ druggist/mechanic/truck driver).
4. A (glasses/pill/watch/lantern) will help you see in the dark.
5. People who are debating an issue are (complaining/arguing over/driving/removing) it.
6. A doctor who specializes in obstetrics is concerned with (eyes/childbearing/broken bones/obesity).
7. When you lose something, you generally look (at it/for it/it up/it over).
8. Another word for *courage* is (silence/inscription/fear/ bravery).
9. A hitchhiker is a person hoping to get a free (ride/meal/ cigarette/watch).
10. If a place is remote, it is (near/beautiful/far/silent).

*Discussion*

1. Have you had any experiences with hitchhikers? Have you ever hitchhiked? What was it like?
2. How do you think the doctor felt after he read the inscription? What lesson could he have learned from this experience?
3. Are you generally suspicious of strangers? Would you like to be more (less) suspicious of people?
4. Are people in rural, remote places more or less trusting of people, in general, than people in big cities? Why?
5. Do you, or does someone in your family, own a gun? Why? What kind? Do you think people should be allowed to own handguns?

# Unit 15: Better Late

*Edward Stevenson*

I had just completed work on my very first screenplay for a major Hollywood studio, and I was exhausted. Being a screenwriter was going to be harder than I thought. I reflected that I had had the hardest time writing dialogue for elderly people. After reading the travel section of *The New York Times* on my first Sunday back in the city, I decided to kill two birds with one stone. I would relax by taking an ocean cruise, and I would purposely sign onto a cruise where there were bound to be a lot of elderly folks. That way I could get some needed rest and also hone my dialogue-writing skills.

I was lucky. On our first day out, I met a fellow who must have been at least seventy-five. He was traveling with his wife, and he was a talker. As best as I can reproduce it, here's what he said to me as we met on the deck of the S.S. Helge, a Norwegian liner.

"Well, I'm certainly glad you're not seasick. When I first saw you leaning over the rail, I said to myself that you must be seasick, though I couldn't see how anybody could get seasick with the water so calm the way it is today. Our room steward says that anybody that gets seasick in this kind of weather wouldn't be safe on the lake in Central Park. He's a regular comedian. . . . And that reminds me, how much do you think I ought to tip him? The room steward, I mean. I'm not a person who has a lot of money, but still I want to do the right thing as to tipping.

"You see, this is the first time we've been on a boat—my wife and I, I mean. Of course, we've taken a trip up the Hudson with the kids, but I guess you wouldn't mention the Hudson River Day Line in the same breath with a big ship like this, would you? The kids thought it was wonderful, though. They're grown up and married now, with kids of their own—except Judy, that is, and she hardly has had time, not having been married a year yet —but it doesn't seem more than yesterday that they were running around and getting into all kinds of trouble. Time certainly flies . . .

"Whew! It's getting hot, isn't it? We must be coming into the tropics from the way it feels. Ever been down here before, Mr. —I don't think you mentioned your name, did you? Arthur? Well, I'm glad to know you, Mr. Arthur. My name's Bentham. I'd like you to meet my wife sometime, too. That's my wife sitting in that deck chair down at the end. She's making believe she's reading that book, but she's sound asleep. The salt air seems to make her very tired . . .

"As I was saying, time certainly flies. Now, you take me; why, it seems only the other day that Ellen and I were getting married; and here we are grandparents of six already.

"We've been married fifty years. It doesn't seem possible, but that's what it is, all right. Why, say, I can remember the wedding just as clearly as if it happened last week. It wasn't much of a wedding—you know, no ceremony and reception. Besides Ellen and me and the minister, there were only the minister's wife and the church janitor for witnesses. I can still see the five of us standing there in the chapel with the sun coming through the stained glass window, falling all around us, turning everything different colors. It was a long time ago, but the memory is still strong.

"Will you look at them flying fish! Aren't they the funniest things!

"When I look back, I think that Ellen and I must have been crazy, getting married the way we did. My goodness, I didn't have a cent to my name—it was all I could do to get together the money for the wedding ring. Engagement rings and honeymoons and all those special things were out of reach as far as we were concerned.

"I felt pretty bad, starting our marriage in a small efficiency apartment over a laundry. I wanted it to be better for us, and I told her so. A big wedding, a reception at the best hotel in New York, and a honeymoon in Miami or the Bahamas. She just laughed. 'If I wanted such riches, I'd have married a Rockefeller and not Johnny Bentham.' That's the way she was. And is. I didn't really mind not having a big church wedding, or a reception afterwards, but, gee, what's a wedding without a honeymoon? It made me feel low.

"But, you know, marrying that woman was the making of me. I was just a shipping clerk at the time; it was during the Great Depression, and I was lucky to have a job at all. Well, Ellen encouraged me to study accounting, and when an opening in the company came along, I stepped right into it. Today, folks call that 'upward mobility'; I called it 'being at the right place at the right time.'

"I retired as comptroller of that company last month. When I left the firm, they put on a dinner for me at the Plaza Hotel and gave me a watch. Here it is. See what it says, 'To John W. Bentham,' that's my full name, 'for fifty years of devoted service.' Mr. Stover, the president, made a speech. I did too—but I was too choked up to say much. You can bet that I'd never have stayed in that company long enough to get that watch if it hadn't been for Ellen.

"And the kids, too, of course. When you get to be a family man you have to be a little more serious. Kit—that's short for Christopher—was the first; then Roger, Cynthia, Anthony, and Judy, the baby. Nice names, aren't they? Ellen picked them out.

"They're all grown up now—fine young men and women, if I do say so myself—but there were times when you just wondered if they ever would grow up. It was just one thing after another. Sick or healthy, they had you up to your neck in bills.

"Is that land over there to the left? No, I guess it's just clouds.

"Well, last year our company did pretty well and they gave all the old employees a month's pay for a bonus at Christmas—first bonus we had had in years. So what did I do? Well, I figured with all the kids married and no one to take care of but ourselves that we didn't have any real need for the money, so I didn't breathe a word about it to Ellen. You see, I'd been seeing those cruise advertisements in the papers and I thought to myself, that's just the thing for Ellen and me. Twelve days. Nassau and Jamaica for $750. I didn't say anything till about two weeks before we were about to sail. Then I broke the news. Well, you could have knocked Ellen over with a feather.

" 'Johnny Bentham,' she says, 'are you out of your mind?'

" 'No,' I says. 'And I haven't robbed a bank, either.' So I told her all about the bonus.

"Well, she still thought I was crazy. 'Spending all that money on a little trip,' she says. 'Do you think we're millionaires? Johnny, I'll never set foot on that boat.'

" 'Now, that's a fine way to feel' I says, acting as if I were insulted. 'A woman refusing to go on a honeymoon with her husband!'

"Well, she just looked at me and I just looked at her, and first thing you knew she threw her arms around me and began kissing me, and what did the two of us do but end up laughing and crying like a couple of kids.

" 'Gee, Mama,' I says. 'It's better late than never, isn't it? . . .'

"Say, look at those flying fish!"

Mr. Bentham and I talked—actually, he talked and I listened—several more times during the cruise. I met Ellen, I met several other people their age, and I relaxed. Who could ask for more than that?

*Comprehension*

1. What does the narrator of the story do for a living? Why is he on the cruise?
2. How long have John and Ellen Bentham been married?
3. How did Bentham describe his wedding? What regrets did he have?

4. Where did the Benthams live right after they were married? How did Bentham feel about this?

5. What was Bentham's job when he got married? What did Ellen encourage him to do?

6. What did Bentham do last month? What position had he held prior to that?

7. How did his company reward him when he retired?

8. Where did the money for the Benthams' cruise come from?

9. How did John overcome Ellen's original objections to taking the cruise?

10. What is the significance of the title of the story?

*Exercises*

**A.** Use each of the following terms in a sentence:
screenplay, to kill two birds with one stone, cruise, to hone, to say to oneself, comedian, to grow up, to run around, to make believe, to be sound asleep, stained glass, honeymoon, efficiency apartment, upward mobility, devoted, bonus.

**B.** Circle the word in parentheses which best completes the sentence.

*Example:* The chief financial officer in a company is called the (clerk/⟨comptroller⟩/screenwriter/steward).

1. You usually give a (bonus/tip/message/watch) to a waiter or waitress.

2. If a body of water is calm, it is (smooth/choppy/rough/windy).

3. Someone who is choked up is about to (laugh/fall asleep/cry/sneeze).

4. A person who is exhausted is very (cold/hot/excited/tired).

5. To make believe is to (cancel/postpone/pretend/instruct).

6. *Sound asleep* means (sleeping lightly/half asleep/sleeping deeply/wide awake).

7. To mind something is to (like it/approve of it/insist on it/object to it).

8. A chapel is a small (ship/hotel/house/church).

9. The people who witness a marriage (avoid/observe/forget/calculate) it.

**10.** An eighty-year-old person is considered (middle-aged/ retired/elderly/late).

**C.** Words like *down, over, out, under,* and *up* are often used as prefixes. Attach one or more of these prefixes to each of the following words. Then use each word in a sentence.

*Example:*   paid   My boss told me I was *overpaid,* but I was actually *underpaid,* so I went out and found a job with a better salary.

| | |
|---|---|
| **1.** put | **6.** fall |
| **2.** charge | **7.** hill |
| **3.** look | **8.** right |
| **4.** line | **9.** town |
| **5.** side | **10.** set |

*Discussion*

**1.** Have you ever taken a cruise? Would you like to take one? Where would (did) you go?

**2.** What does a shipping clerk do? A comptroller?

**3.** What do you think of the concept of luck, or "being at the right place at the right time"? Are you a lucky person?

**4.** Do you know any people who have been married for fifty years? What are they like?

**5.** What is a depression? What do you know about the Great Depression which Bentham referred to? When was it? How did it affect your family?